SpringerBriefs in Environmental Science

SpringerBriefs in Environmental Science present concise summaries of cutting edge research and practical applications across a wide spectrum of environmental fields, with fast turnaround time to publication. Featuring compact volumes of 50 to 125 pages, the series covers a range of content from professional to academic. Monographs of new material are considered for the SpringerBriefs in Environmental Science series.

Typical topics might include: a timely report of state-of-the-art analytical techniques, a bridge between new research results, as published in journal articles and a contextual literature review, a snapshot of a hot or emerging topic, an in-depth case study or technical example, a presentation of core concepts that students must understand in order to make independent contributions, best practices or protocols to be followed, a series of short case studies/debates highlighting a specific angle.

SpringerBriefs in Environmental Science allow authors to present their ideas and readers to absorb them with minimal time investment. Both solicited and unsolicited manuscripts are considered for publication.

More information about this series at http://www.springer.com/series/8868

Bo Qu

The Impact of Melting Ice on the Ecosystems in Greenland Sea

Correlations on Ice Cover, Phytoplankton Biomass, AOD and PAR

 Springer

Bo Qu
School of Science
Nantong University
Nantong, Jiangsu
China

ISSN 2191-5547   ISSN 2191-5555   (electronic)
ISBN 978-3-642-54497-2 ISBN 978-3-642-54498-9   (eBook)
DOI 10.1007/978-3-642-54498-9

Library of Congress Control Number: 2014949369

Springer Heidelberg New York Dordrecht London

This work was supported by the National Science Foundation of China (Grant No. 41276097).

Printed on acid-free paper

Springer is part of Springer Science+Business Media (www.springer.com)

Preface

It has been argued that the Arctic is a sensitive indicator of global change. The ice cover in Arctic Ocean provides a control not only on the surface heat and mass budgets of the Arctic Ocean but also on the global heat sink. It has also been suggested that an enhanced pool of Arctic and freshwater on the ocean surface coming from melting ice may significantly affect the global ocean thermohaline circulation. Changes in sea-ice cover will affect not only the physical Arctic Ocean, but also result in chemical, biological, and ecosystem changes. The impact of melting ice on oceanic phytoplankton and climate forcings in the Arctic Ocean has attracted increasing attention due to its special geographical position and potential susceptibility to global warming.

Salty sea smell near the ocean does not result from the salt alone. Gases diffuse across the air-sea interface, many of which are synthesized and emitted by microalgae. One of these gases is a sulfur-based compound that has a strong characteristic odor. It has been suggested that variations in algal production of these natural gases play an important role in moderating our climate through their aerosol effect on backscattering solar radiation and in cloud formation. Scientists have identified the sulfurous gas as dimethylsulfide (DMS). DMS is a naturally produced biogenic gas essential for the Earth's biogeochemical cycles.

In the ocean, DMS is produced through a web of biological interactions. Certain species of phytoplankton, microscopic algae in the upper ocean synthesize the molecule dimethylsulfoniopropionate (DMSP), which is a precursor to DMS. When phytoplankton cells are damaged, they release their contents into the seawater. Bacteria and phytoplankton are involved in degrading the released algal sulfurous compound DMSP to DMS and other products. A portion of the DMS diffuses from saltwater to the atmosphere. Once it is transferred to the atmosphere the gaseous DMS is oxidized to sulfate aerosols, and these particulate aerosols act as cloud condensation nuclei (CCN) attracting molecules of water. Water vapor condenses on these CCN particles forming the water droplets that make up clouds. Clouds affect the Earth's radiation balance and greatly influence regional temperature and climate. DMS represents 95 % of the natural marine flux of sulfur gases to the atmosphere, and scientists estimate that the flux of marine DMS supplies about 50 % of the global biogenic source of sulfur to the atmosphere. Greenhouse gases have well-constrained positive forcings (creating a warming). In contrast, DMS air-sea fluxes have negative forces creating a cooling effect.

At its maximal extent, sea-ice covers over 80 % of the Arctic Ocean. Sea-ice plays a dominant role in determining the intensity of the DMS fluxes in the Arctic and the Antarctic and to a large extent determines the climate sensitivity of both regions. The decline in sea-ice cover would have an effect on phytoplankton dynamics and ocean circulation systems and hence have a significant impact on the global climate.

Here I studied the sea-ice impact on the Greenland Sea ecosystem. Greenland Sea is located on the west of the Arctic Ocean and east of Greenland where the world's second largest glaciers are located. The sea-ice has great impact on the local phytoplankton communities. The correlation study is essential for the overview of the local ecosystem. The analysis results and methods provided here not only give an outline of the ecosystem in Greenland Sea in the recent decade and how the ice impacts the local ecosystems, but also provide valuable statistical methods on analyzing correlations and predicting the future ecosystems.

As a research fellow, I worked in Griffith University, Brisbane from 2003 to 2006. I worked for a project of the biogeochemistry research in Arctic Ocean undertaken by Prof. Albert Gabric, a well-known DMS modeling expert in the world. We carried out ecosystem research in Barents Sea. It is found that temporal and spatial distribution of phytoplankton biomass (measured using chlorophyll-a (CHL)) is strongly influenced by sea-ice cover, light regime, mixed layer depth, and wind speed in Barents Sea. Later, we used genetic algorithms to calibrate a DMS model in the Arctic Ocean. The general circulation model (CSIRO Mk3) was applied to calibrate DMS model to predict the zonal mean sea-to-air flux of DMS for contemporary and enhanced greenhouse conditions at 70 °N–80 °N. We found that significant ice cover decrease, sea surface temperature increase, and mixed layer depth decrease could lead to annual DMS flux increases by more than 100 % by the time of equivalent CO_2 tripling (the year 2080). This significant perturbation in the aerosol climate could have a large impact on the regional Arctic heat budget and consequences for global warming. Leon Rotstayn, the Principal Research Scientist from Marine and Atmospheric Research Centre in CSIRO supervised the GCM batch system running.

The cooperation research with Australia has been carried on since then. My Chinese national natural science funding entitled "The Impact of Arctic Ecosystem and DMS to its Climate" provided us with further research possibilities.

Sincere thanks should first go to Prof. Albert Gabric for his opening the door and leading to the further study of this project. Huge thanks to my four students: Li Hehe, Gu PeiJuan, Dong LiHua, and Wang ZaQin, for their hard work on processing regional satellite data. Great thanks to Chinese national natural science funding for providing the possibilities on carrying work on the project.

Nantong, August 2014   Bo Qu

Acknowledgments

Huge thanks should go to my four students: Li HeHe, Gu PeiJuan, Dong LiHua and Wang ZaQin. Thanks for their hard work on getting those regional satellite data and processing them as well. Sincere thanks to my previous supervisor Prof. Albert Gabric in Griffith University, Australia, for his guidance and leading me to this Arctic ecosystem research area. Thanks to NASA's Ocean Biology Processing Group for providing MODIS aqua, Level 3 (4-km equi-rectangular projection) 8-day mapped data for chlorophyll-*a* (CHL) and aerosol optical depth (AOD). Thanks to NASA Goddard Space Flight Centre of SeaWiFS Project group for providing 8-day mapped Photosynthetically Active Radiation data. Thanks to NASA Web SeaDAS development group for providing Ocean Colour SeaDAS Software (SeaWiFS Data Analysis System) for processing CHL, AOD, and PAR data. Thanks to NOAA NCEP EMC CMB GLOBAL Reyn-SmithOIv2 for providing weekly and monthly sea-ice concentration. Thanks to NASA for providing Wind Data and Sea Surface Temperature data. WindSat data are produced by Remote Sensing Systems and sponsored by the NASA Earth Science MEaSUREs DISCOVER Project and the NASA Earth Science Physical Oceanography Program. Thanks to NASA http://gdata1.sci.gsfc.nasa.gov for providing cloud cover data.

Thanks to the Chinese National Natural Science Funding (Funding No. 41276097) for providing funding for this project. Thanks to the Chief Editor Lisa Fan and other editors, for all the initiation and hard work toward getting this book organized until publishing.

Contents

Chapter 1
Overview Greenland Sea

Abstract Arctic marine ecosystems are largely impacted by global warming. The sea ice in Greenland Sea plays an important role in regional climate system and even to the global climate changing. The special characters of the surface current in Greenland Sea are outlined. The melting ice (MI) effect on the climate system is emphasized. The relationships between North Atlantic Oscillation (NAO) and ice cover (ICE) for different situations are also listed. Finally, the important roles of sea ice on the ecosystem for different aspects are described.

Keywords Greenland Sea · Ice cover (ICE) · Melting ice (MI) · Current · NAO

Greenland Sea is located in the southeast of Arctic Ocean. Its marine environment and ocean circulations are highly dominated by North Atlantic Ocean. To its west is the Greenland with the world's second largest glacier. Only the fjords areas (near shore) are dominated by local conditions (river runoff, ice formation, etc.). It is a highly dynamic area for water mass exchange between North Atlantic water from south and the Arctic water from north. It is also the area where most Arctic drifting ice is advected. Hence, Greenland Sea is the best region for studying the relationship between MI and phytoplankton biomass in the world. The most in situ and satellite chlorophyll data are also available in this area (Arrigo and van Dijken 2011).

1.1 Current

Surface current is shown in Fig. 1.1 around Greenland Sea (including part of Iceland Sea and Norwegian Sea). East Greenland Current (EGC) moves from north to south along east Greenland coastline, brings colder, less saline Arctic water to southern ocean. In south of Iceland, the current is from warmer more saline southwest of north Atlantic, along Norwegian current all the way up to north into Arctic ocean. Between 70°N and 80°N, there is an anticlockwise

B. Qu, *The Impact of Melting Ice on the Ecosystems in Greenland Sea*,
SpringerBriefs in Environmental Science, DOI 10.1007/978-3-642-54498-9_1

Fig. 1.1 Surface current and study region surrounding Greenland Sea

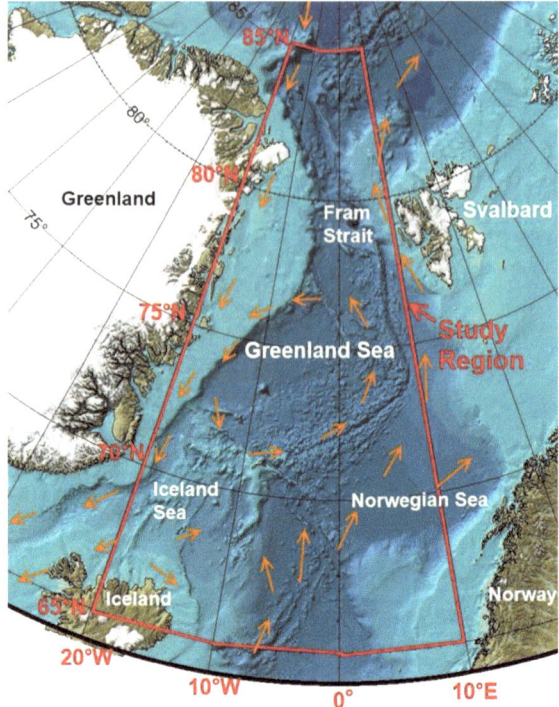

direction swirling current in the study region. In this region, the vertical stability increased to the north by the input of meltwater and solar heating, phytoplankton biomass would increase, and nutrients concentration would decrease in the region (Lara et al. 1994). At around 70°N, the southward Atlantic water splits into two parts: one part along the west coast of Norway flows into Barents Sea on its east, and the other part northward to the Spitsbergen region. The polar front is located along the east of East Greenland Current, and Arctic front is located along the west of Norwegian Current.

1.2 The MI Effect

The significant decline of Arctic sea ice resulted in the rising of sea-ice level and temperature especially in Arctic Ocean. It was reported that the Arctic temperature has increased at twice the rate as the rest of the globe and could increase continuously by the end of this century. The Arctic sea ice helps to regulate global temperature by reflecting sunlight back into space. With the large area of ice loss, replacing bright sea ice with dark ocean (it absorbs sunlight), it is the region for speeding global warming. Because the Arctic sea-ice extent decreased by 12 % per decade, the Arctic autumn air temperature has increased by 4–6 °F in the past decade (http://www.wunderground.com/climate/SeaIce.asp).

The direct effect of melting of Arctic sea ice is the change of sea level. The global sea level would rise to about 4 mm if all world sea ice was melted. Greenland's ice added 6 times more to sea levels in the decade in the previous 10 years, according to a draft of the UN's most comprehensive study on climate change. The indirect effect of melting sea ice is the warmer average temperatures locally and globally. Warmer temperatures will accelerate the melting of the Greenland ice sheet, which holds enough water to raise sea level 20 feet. The Arctic ice retreated extensively, and the first-year ice is thinning, that is, vulnerable to more summer melting. Arctic ice could be totally gone by 2030 (Stroeve et al. 2007).

1.3 The Arctic Amplification and NAO

The Arctic sea ice is an important indicator of the global climate system. The major reason is that the ice could regulate heat exchange between relative colder atmosphere and warmer ice-covered ocean in winter (Jaiser et al. 2012). The temperature rising in Arctic is much larger than Northern Hemisphere or the globe as a whole. The phenomenon is called the Arctic amplification (Serreze and Barry 2011).

It was found that the recent Arctic amplification is much more significant in autumn and winter seasons and is much weaker in spring and summer seasons (Screen and Simmonds 2010). Hurrell (1995) pointed out that the warmer temperature in winter over Eurasia indicated the positive tendency of North Atlantic Oscillation (NAO) in winter. However, the negative NAO winter value indicated decline of sea ice in North Atlantic and more ice melting in Greenland (Jaiser et al. 2012). Autumn and warming patterns were found to associate with the reduction in ice cover in September (Serreze et al. 2009). Screen and Simmonds (2010) also confirmed the Arctic amplification in recent years is due to the reduction in sea-ice extent in September. As the ice melts, the open water area would expose and would absorb the solar radiation, and hence would increase the surface water temperature and shallow the mixed layer depth. This will lead to further ice melting. When Arctic sun ends (in the beginning of autumn), there would still be larger heat transfers from the ocean to atmosphere. Hence, the autumn warming occurs.

1.4 Sea-Ice Ecosystem

The presence of sea ice affects a wild range of important processes: light transmitting, heat, and gas exchange and stability of the water columns. With more MI, more diluted water column could be formed underneath. During spring and early summer, when temperatures begin to increase, ice algal communities dominated by diatom would appear. Sea ice cover could influence phytoplankton blooms by reducing light penetration to the water column, and hence reduce the growth rate of algae in/under the sea ice. During sea-ice melting, sea-ice plankton, nutrients, and trace elements are released to the upper water layer; it would accelerate the bloom process (Cherkasheva et al. 2014). Moreover, the MI added more freshwater into Upper Ocean and could increase the stability of the surface water. When light is favorable, it would promote blooms. On the other hand, it could suppress the bloom by increasing grazing pressure from zooplankton or limit nutrients supply from deeper layers and thus constrain the growth of blooms (Cherkasheva et al. 2014).

Sea-ice ecosystem provides food for variety of animals. The decrease in sea ice would threaten the local animals, and hence destroy the ecosystems.

Sea-ice algae could be exposed to a wide range of light conditions. Hence, low-light winter also could produce ice algae. Cui et al. (2012) found that phytoplankton communities in fjords (Spitsbergen in northeast of Greenland Sea) are darkness adapted in late summer. The Greenland glacial meltwater is favorable for phytoplankton growth when influence of freshwater is limited. Arrigo et al. (2014) found that high light and UV inhibit photosynthesis in sea-ice diatoms by diminishing photosystem performance and photosynthetic rate and increasing DNA damage rate. Dimethylsulfoniopropionate (DMSP), the precursor of DMS, produced by sea-ice diatoms, would have important feedback on the climate systems via production of sulfate aerosols, which are important cloud nucleation surfaces (Charlson et al. 1987).

Nutrient limitation would prevent ice algae growth. The situation would happen in low temperature (winter) and high salinity. Salinity of 30 favors algae growth, neither too low nor too high (Arrigo et al. 2014). Cui et al. (2012) also found the high diluted water had negative influence on phytoplankton growth. Arrigo et al. (2014) pointed out that the silicic acid is the macronutrient, which limits the algal growth. Iron as the main micronutrient usually concentrated in sea ice and is generally in ample supply.

Bacteria populations would increase dramatically through spring and summer, response to increase organic carbon supplies for growth of algal blooms. The carbon is transferred from bacteria to phages and protists. Protists play an important role in controlling bacterial populations (Delille et al. 2002). Sea ice is a more favorable bacterial habitat than water column (Martin et al. 2010).

With sea-ice algal blooms reached to its peak in spring and early summer, the availability of light in upper water column would reduce. The water column blooms of phytoplankton would delay until ice algal bloom has subsided (Arrigo et al. 1991).

The biogeochemical processes in sea ice involve the following processes: macronutrients, trace elements, organic and inorganic carbons, other gases (such as DMS, methane, etc.), and atmospheric halogen chemistry, with strong interaction with oceanic and atmospheric processes (Vancoppenolle et al. 2013). The key mechanisms that determine phytoplankton growth in Greenland Sea are ice cover, stratification, wind, surface transport, and the activity of grazers (Slagstad et al. 2011). Hence, research on impact of ice melting on ecosystem especially on phytoplankton growth is a complicated task. Here, only the correlations among ice cover, Chlorophyll-a, aerosol optical depth, and North Atlantic Oscillation are investigated. Some predictions would be done later in the book.

References

Arrigo, K. R. (2014). Sea-ice ecosystems. *Annual Review of Marine Science, 6*, 439–467. doi:10.1146/annurev-marine-010213-135103.

Arrigo, K. R., Sullivan, C. W., & Kremer, J. N. (1991). A bio-optical model of Antarctic sea-ice. *Journal Geophysical Research, 96*(C6), 1058192.

Arrigo, K. R., & van Dijken, G. L. (2011). Secular trends in Arctic Ocean net primary production. *Journal of Geophys. Res., 116*, C09011.

Charlson, R. J., Lovelock, J. E., Andreae, M. O., & Warren, S. G. (1987). Oceanic phytoplankton, atmospheric sulphur, cloud albedo and climate. *Nature, 326*, 655–661.

Cherkasheva, A., Nöthig, E. M., Bauerfeind, E., Melsheimer, C., & Bracher, A. (2014). From the chlorophyll-a in the surface layer to its vertical profile: a Greenland Sea relationship for satellite applications. *Ocean Science, 9*, 431–445.

Cui, S., He, J., He, P., Zhang, F., Lin, L., & Ma, Y. (2012). The adaptation of Arctic phytoplankton to low light and salinity in Kongsfjorden (Spitsbergen). *Advances in Polar Science, 23*, 19–24.

Delille, D., Fiala, M., Kuparinen, J., Kuosa, H., & Plessis, C. (2002). Seasonal changes in microbial biomass in the first-year ice of the Terre Adelié area (Antarctica). *Aquatic Microbial Ecology, 28*, 257–265.

Hurrell, J. W. (1995). Decadal Trends in the North Atlantic Oscillation: Regional Temperatures and Precipitation. *Science, 269*, 676–679.

Jaiser, R., Dethloff, K., Handorf, D., Rinke, A., & Cohen, J. (2012) Impact of sea-ice cover changes on the Northern Hemisphere atmospheric winter circulation. *Tellus A 64*.

Lara, R. J., KATTNER, G., Tillmann, U., & Hirche, H. J. (1994). The North East Water polynya (Greenland Sea) II. Mechanisms of nutrient supply and influence on phytoplankton distribution. *Polar Biology, 14*, 483–490.

Martin J, Tremblay JÉ, Gagnon J, Tremblay G and others (2010) Prevalence, structure and properties of subsurface chlorophyll maxima in Canadian Arctic waters. Mar Ecol Prog Ser 412:69-84.

Serreze, M. C., Barrett, A. P., Stroeve, J. C., Kindig, D. N., & Holland, M. M. (2009). The emergence of surface-based Arctic amplification. *The Cryosphere, 3*, 11–19.

Serreze, M., & Barry, R. (2011). Processes and impacts of Arctic Amplification Global and Planetary Change:A research synthesis. *Global and Planetary Change, 77*, 85–96.

Slagstad, D., Ellingsen, I. H., & Wassmann, P. (2011). Evaluating primary and secondary production in an Arctic Ocean void of summer sea-ice: An experimental simulation approach. *Progress in Oceanography, 90*, 117–131.

Stroeve, J., Holland, M. M., Meier, W., Scambos, T., & Serreze, M. (2007). Arctic sea-ice decline: Faster than forecast. *Geophysical Research Letters, 34*, L09501.

Vancoppenolle, M., Bopp, L., Madec, G., Dunne, J., Ilyina, T., Halloran, P. R., et al. (2013). Future Arctic Ocean primary productivity from CMIP5 simulations: Uncertain outcome, but consistent mechanisms. *Global Biogeochemical Cycles, 27*(3), 605–619.

Chapter 2
Chlorophyll a, Ice Cover, and North Atlantic Oscillation

Abstract This chapter investigated the relationships between phytoplankton biomass, measured using chlorophyll *a* (CHL), sea-ice cover (ICE), and North Atlantic Oscillation (NAO) in the Greenland Sea in 20°W–10°E, 65–85°N during the period 2003–2012. Remote-sensed satellite data were used to do correlation analysis. Enhanced statistics methods (such as unit root checking, lag regression, and co-integration analysis methods) are used for correlation analysis. Results show that the melting ice (MI) played a significant role on promoting the growth of CHL. In general, ICE reached peak (in March) 3 months ahead of CHL (peaked in June), and CHL was higher in south and lower in north. CHL increased around 10 % in spring and early summer during last 10 years in 75°N–80°N. Moreover, CHL was higher in 75°N–80°N region where ice melted more and the water column was more stable. The peak of CHL in 2012 was 1 month later than the other years. The CHL peak in 2011 was highest, and there were two peaks in 2010. The peaks of CHL came later in 2012 and 2008. The early and higher peaks of CHL in year 2010 was due to the more MI happened in that year, Other reasons including the stronger wind speed in spring and special wind direction from southeast changed to southwest, plus lower SST and PAR in summer and negative NAO through the year. The research shows that CHL, ICE, and NAO were correlated with a time lag. CHL and ICE had long-term equilibrium relationship. The NAO and MI had a negative correlation. NAO affected the MI and its peak was 3 months ahead of the MI. The CHL and NAO also had negative correlations. With NAO reached to its peak, CHL almost reached to its valley at the same time.

Keywords Chlorophyll a (CHL) · Ice cover (ICE) · Melting ice (MI) · North Atlantic Oscillation (NAO) · Peak · Coupling

© The Author(s) 2015
B. Qu, *The Impact of Melting Ice on the Ecosystems in Greenland Sea*,
SpringerBriefs in Environmental Science, DOI 10.1007/978-3-642-54498-9_2

2.1 Introduction

2.1.1 Sea Ice and the Phytoplankton Biomass

Sea ice provided a significant amount of habitat for productive microbial communities (including algae, bacteria, archaea, heterotrophic protists, and viruses) (Horner et al. 1992). In terms of biomass, the communities were dominated by algae, particularly diatoms during bloom period (Vancoppenolle et al. 2013). There are many protist species in Arctic sea ice, with diatoms dominated, other species such as dinoflagellates and chrysophytes. Ice algae also provided early-season high-quality food source for pelagic herbivores (Soreide et al. 2010).

There are numerous studies on the MI and its contribution to the phytoplankton concentrations (Matrai and Vernet 1997; Wassmann et al. 1999; Olli et al. 2002; Qu et al. 2006; Pabi et al. 2008; Leu et al. 2011). It was suggested that the decreasing of sea ice and increasing of light result in increasing of phytoplankton biomass. What is the effect of phytoplankton to ice cover properties? Early study from Ericken et al. (1991) suggested that high phytoplankton biomass may accompany with low ice strengths. The reason is ice algae may speed ice deterioration and increase porosity via solar radiation absorption.

2.1.2 The Light Effect on Phytoplankton Biomass

Phytoplankton would decrease with the increase of light intensity in summer. There is an optimal light intensity for growth of phytoplankton. Algae under ice receive much less light than in open area. The question is, how much light is least required for growth of phytoplankton and how much nutrients required as well? Jassby and Platt (1976) obtained a mathematical formula of the relationship between photosynthesis and light for phytoplankton. They derived the following formula:

$$P^B = \alpha I - R^B \tag{1.1}$$

where P^B is the primary production per unit chlorophyll biomass, I is irradiance, and α is the slope of the light-saturation curve at low light levels. Light-independent respiration loss R^B ($mgC[mgChla]^{-1}h^{-1}$) is

$$R^B = P^B_{gross} - P^B_{net} \tag{1.2}$$

It is interesting to know that the phytoplankton in Arctic Ocean can survive without irradiance (Parsons et al. 1984). Melting water generates more nutrients, although the dilution decreased salinity and surface temperature and also brings lower light penetration, these could have negative effect on the growth of phytoplankton (Cui et al. 2012).

In Arctic Ocean, light is very important factor controlling phytoplankton biomass. Compared to the ice-covered region, polynyas received much earlier light in the year. Hence, earlier CHL appeared in the polynyas region. The Arctic water flows from northeast and formed upper layer waters, while North Atlantic water flows from South and into deep layer waters. With the phytoplankton biomass increased, the nutrient concentrations decreased. Lara et al. (1994) found that diatom appeared often in open water and in the starting production period. In late spring (April) in northern part of Greenland Sea, most of species forming the spring bloom are located under the ice. They are both diatoms and flagellates. Phytoplankton biomass growth until nutrients depleted. Phytoplankton advected from north to south by anticyclonic pattern (Schneider and Budeus 1994).

The different stages of ice melting would add different amount of ice algae to the community. The process is complicated due to many effects. The detailed study on ice melting and its relationship with phytoplankton biomass and other effects of decline ice is expected to carry out. This chapter is focused on the effect of ice melting on the phytoplankton biomass (CHL) and their relationship with NAO based on the most recent 10 years data in Greenland Sea.

2.2 Data and Methods

Our study region is in Greenland Sea 20°W–10°E, 65–85°N (highlighted box in Fig. 1.1), for the period of 10 years: 2003–2012. Due to the special condition in Arctic (half year darkness from October to February) and satellite data only valid within March to September, we choose MODIS satellite afternoon (Aqua), 8-day, 4-km, level-3 mapped data for retrieving global chlorophyll a (CHL), aerosol optical depth (AOD) data. MODIS Web site is located in http://modis.gsfc.nasa.gov/. Photosynthetically active radiation (PAR) was derived from SeaWiFs, 8-day mapped data (http://oceandata.sci.gsfc.nasa. gov/seawifs). The image analysis package SeaWiFS Data Analysis System (SeaDAS 6.4) (http://seadas.gsfc.nasa.gov/) was then used to get subset data for our focused study region.

Global sea ice weekly data were obtained from NOAA (ftp://sidads.colorado.edu/ pub/DATASETS). Ice cover is calculated from http://iridl.ldeo.columbia.edu/ SOURCES/ Wind speed, wind directions, and sea surface temperature (SST) were calculated from www.remss.com/windsat. Daily data were downloaded for calculating weekly and monthly mean. Cloud cover (CLD) is from http://gdata1.sci.gsfc. nasa.gov/daac-bin/G3/.

Enhanced statistics methods, such as lag regression method and co-integration analysis method, are used for correlation analysis and long-term equilibrium relationship between two variables.

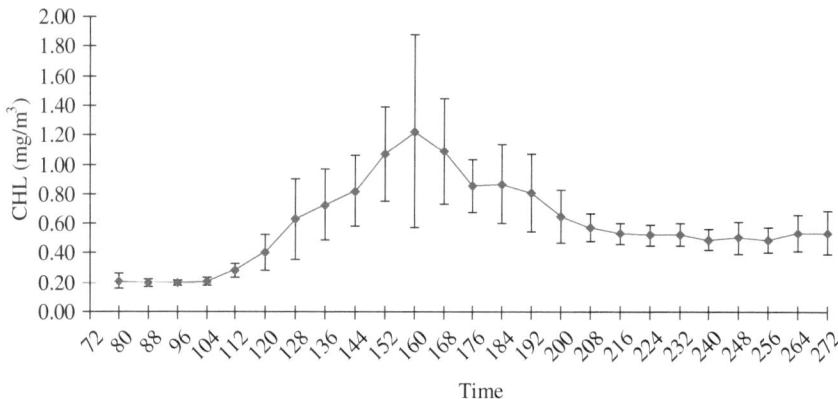

Fig. 2.1   Mean CHL for 8-day interval in the study region averaged for 2003–2012

2.3 Results

2.3.1 CHL Distributions

8-day mean time series of CHL in the study region averaged for the 10 years (2003–2012) is shown in Fig. 2.1. With gradually increased CHL from March and reached to peak in day 160 (early May), then decreased toward the end of September, there was little lump after June.

Looking in detail for different years, we divide the region into 4 subregions with 5-degree zonal difference for each subregion (Fig. 2.2).

In 65°N–70°N (Fig. 2.2a), year 2011 showed the early peak around day 128 (early April) although some missing values after day 128. Year 2003 had the highest peak on day 136 (middle of April) and the second peak was on day 152 (early May). The Russian 2003 fire could be the cause (Serreze et al. 2000). Further north, there was no such high peak in year 2003. Year 2010 had longer peak period from day 144–152 (late April–early May), and highest autumn peak on day 232 (late July). Year 2006 had late peak on day 192 (middle of June).

In 70°N–75°N (Fig. 2.2b), year 2007 had highest peak around day 160 and 2006 had second high peak around day 168 (mid of May). Year 2011 had early rising in April but had some missing data after middle of April. It had the third highest peak around day 160. Year 2008 had double peaks in day 152 and day 168, while 2012 had the latest peak in day 192 (middle of June). The CHL reduced greatly after day 208 (early July).

In 75°N–80°N (Fig. 2.2c), CHL had early peak (in day 128, middle of April) in year 2010 and also had second even higher peak (day 184, early June). Year 2011 had the highest peak in day 160, while 2008 and 2012 had late peak in day 192 (middle of June) with year 2008 much higher than year 2012. We noticed with the

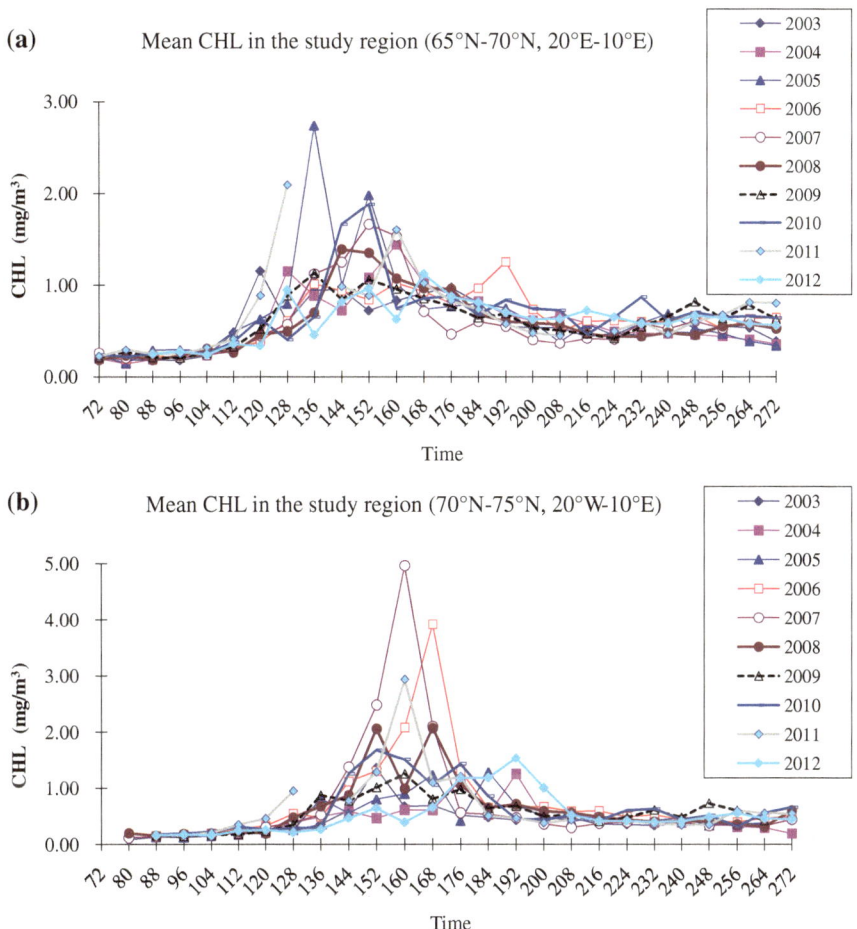

Fig. 2.2 Mean CHL in different subregions for years 2003–2012

latest peak in 2012 and lasted for 8 more days, it had highest autumn peak in day 240 (early August).

In 80°N–85°N (Fig. 2.2d), due to its darkness, CHL started later than southern regions. The satellite data are only valid from April to August. The highest peak was in year 2008 (day 224, middle of July), and year 2004 had early peak in day 136, followed the second peak in day 160–168, then the third peak was in day 224 (middle of July). Year 2006 had early rising of CHL (day 120 early April) and reached to second peak on day 136 and then decreased significantly, increased sharply to reach to its peak on day 152. The rising of CHL in spring of 2006 is interesting and could relate to the patterns of MI. Day 224 had several peaks for years 2003, 2004, 2005, 2008, and 2012. The reasons are worth to find out.

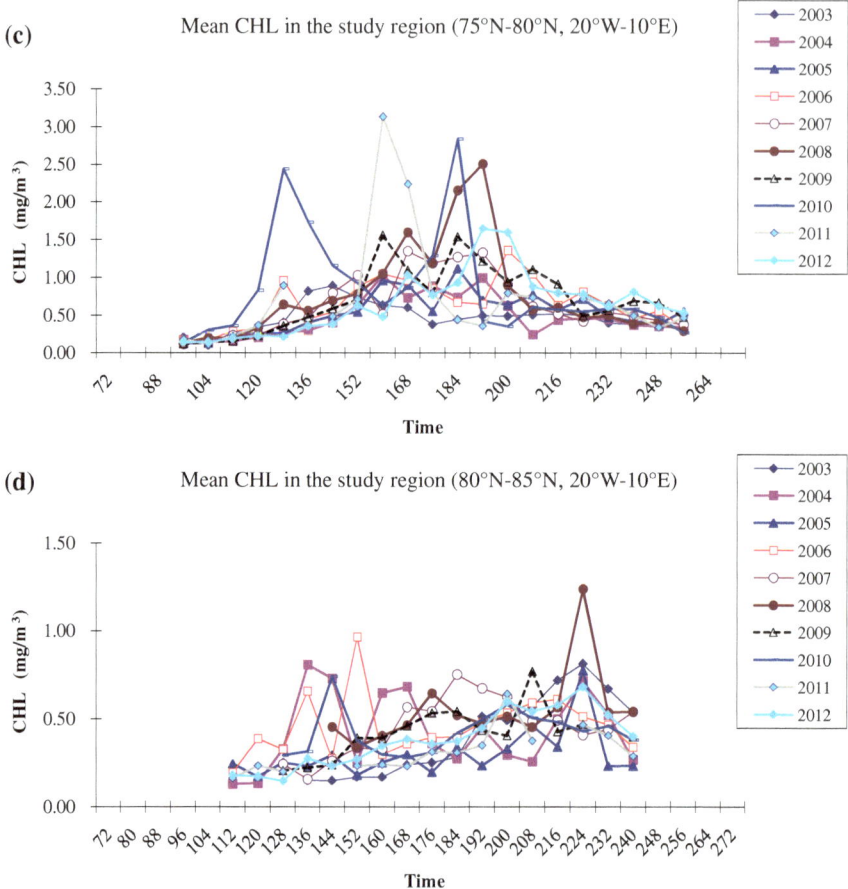

Fig. 2.2 (continued)

Table 2.1 The average CHL peak time during years 2003–2012 for different subregions

	80°N–85°N	75°N–80°N	70°N–75°N	65°N–70°N
Peak time (day)	224	184	160	152

The general trends of CHL increased from north to south, up to 70°N. However, CHL in 65°N–70°N was lower than 70°N–75°N, as there was no ice cover in 65°N–70°N. The average peak time was shifted ahead from day 152 in the south to day 224 in the north (Table 2.1). The time lag was about two and half months.

The detailed peak times for different years in the different subregions are also calculated (Table 2.2). In southern region 65°N–70°N, CHL was gradually shifted ahead from year 2006. In other subregions, CHL peak times in years 2012 and 2008 were much later than other years. However, the first peak time in year 2010 was much earlier. In 75°N–80°N, years 2006, 2008, and 2012 had much late peak time.

Table 2.2 CHL peak time in different years and different subregions

	65°N–70°N	70°N–75°N	75°N–80°N	80°N–85°N
2003	120	152	144	224
2004	160	192	160	136
2005	136	184	184	224
2006	192	168	200	152
2007	152	160	168	184
2008	144	152	192	224
2009	136	160	160	208
2010	152	152	184	144
2011	128	160	160	200
2012	128	192	192	224

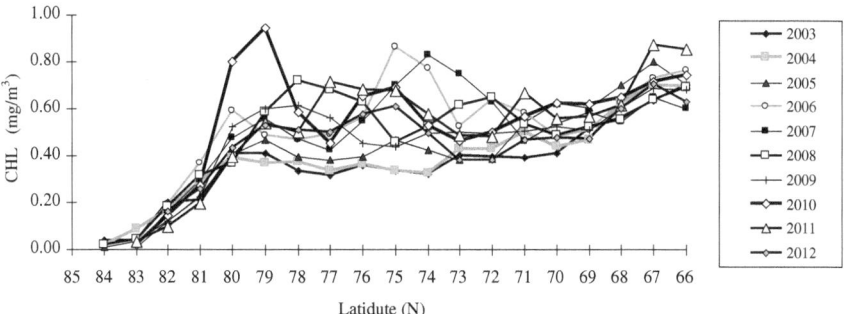

Fig. 2.3 CHL variation along latitude for years 2003–2012

2.3.2 The Reason of the High CHL Peaks in Northern Region

Figures 2.3 and 2.4 are the mean CHL along latitude and longitude for the 10 years. Generally, CHL was higher down south and lower up north. Year 2010 had unusual high peak near 79°N, the magnitude was even greater than southern region. Along longitude, CHL was also higher in year 2010, especially between 18°W–12°W and 2°E–8°E, where East Greenland current and West Greenland current plus Norwegian Sea current located. Year 2004 had least CHL along latitude and year 2003 had least CHL along longitude. Generally, CHL has less variability along longitude for each year.

Figure 2.5 is the mean distribution of CHL in the study region in year 2010. The peak value appeared near 79°N. It is unusual that summer peaks in northern region (near 80°N) were even lower than spring peak.

Generally, CHL distributed higher down south and lower up north in Arctic Ocean (Qu et al. 2006, 2014). However, in our study region, there was a high peak

Mean CHL along longitude in the study region (65°N-85°,N 20°W-10°E)

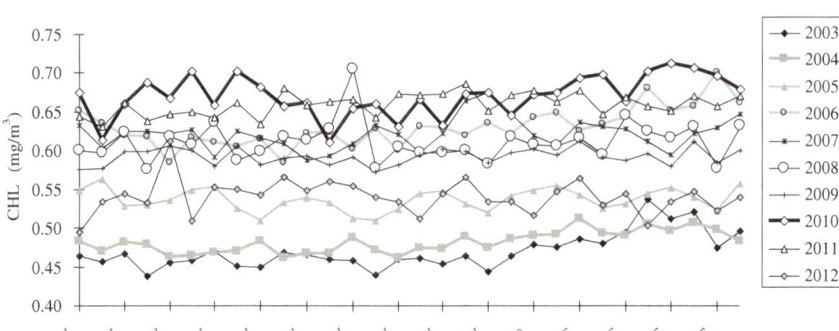

Fig. 2.4 CHL variation along longitude for years 2003–2012

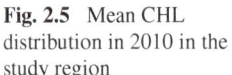
Fig. 2.5 Mean CHL distribution in 2010 in the study region

Mean CHL (mg/m³) in 2010 in the study region

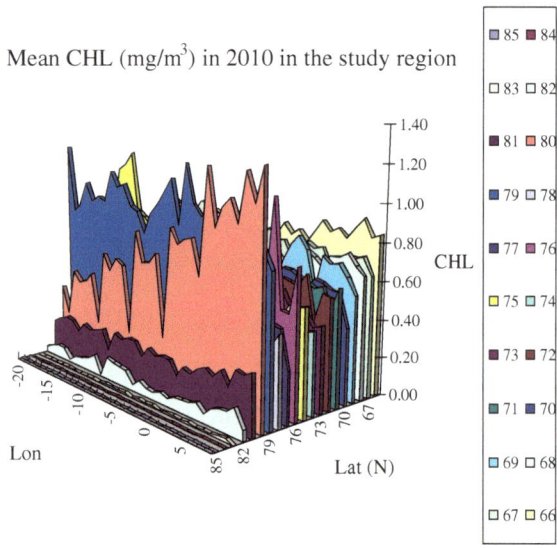

of CHL near 79°N. The reasons causing such a higher northerly peak are unusual. We need to look into several factors in the region.

Lara et al. (1994) did detailed research on the mechanisms of nutrients supply and influence factors on phytoplankton distribution in northeast water of Greenland Sea (78°N–82°N, 20°W–0). They found the vertical stability in their study region was much better than south of the study region. The reason could be the input of melting water and solar heating. The melting water caused lower salinity and higher CHL production. The salinity was lower near 79°N–80°N and higher in both south of 79°N and north of 80°N. Hence, more melting waters from various sites and the land runoff from east of Greenland are the two main factors. Those introduced vertical stability and more iron content, which favored phytoplankton biomass.

Several researchers studied the phytoplankton density near Fram Strait in Greenland Sea and found that the phytoplankton biomass in northeastern of Fram Strait (78°N–81°N) was higher due to the enhanced water–column stability (Gradinger and Baumann 1991). Cherkasheva et al. (2014) did study on Fram Strait area (76°N–84°N, 25°W–15°E). They found the late ice retreat leads to a late ice-associate bloom in the northern region. The stratification of the surface water due to solar radiation (considered is the first reason) and ice melting (the second reason) in the relative sallower surface layer is correspondent to the highest CHL. The water salinity was not much related to CHL. Here, the key parameter for surface stratification is the surface temperature.

Cherkasheva et al. (2014) found that there is no significant relationship between the stratification and CHL variability in coastal water. In coastal water, CHL is higher when absence of ice. This indicating CHL related more to the nutrients rather than light limitation in coastal water. NAO, air temperature, and wind speed could have more impact on marine organism productivity. They also found the phytoplankton blooms would start when the depth of the stratified layer is at its maximum. The later summer months in Fram Strait, CHL concentration decreasing could be due to the limitation of light, stratification, and intense grazing pressure by small copepods and protozooplankton.

The surface melting water south of 79°N and north of 80°N may be depleted from nutrients and lacked vertical stability in the water column due to the different geographic positions.

2.3.3 Ice Cover Distributions

The profile of mean ICE in the 10 years in the study region was generally higher in March and decreased through summer and reached to the valley in September and then increased again after September (Fig. 2.6). Figure 2.7 shows the mean ICE in different subregion (20°W–10°E). There was a dip in year 2009 in spring in northern regions (Fig. 2.7c, d). More ICE happened in spring and summer of 2012 down south. Higher ICE occurred in spring and early summer in 2010 down south. Less ICE occurred in 2004 and 2003 in late summer and early autumn up north.

Figure 2.8 is MI in 75°N–80°N region. We calculated the MI by subtract the ice cover from this week to last week. We are more interested in those higher CHL peak times (Fig. 2.3).

The first CHL peak of 2010 happened in middle of April, while MI started increasing (blue line). The early high March melting in that year contributed relative amount of ice algae to this peak. The second peak of 2010 also happened when MI increased. However, the timing of the two 2010 CHL peaks all happened only one more weeks after MI increasing. The further melting of ice did not contribute more ice algae to the plankton production. The previous MI could contribute to its

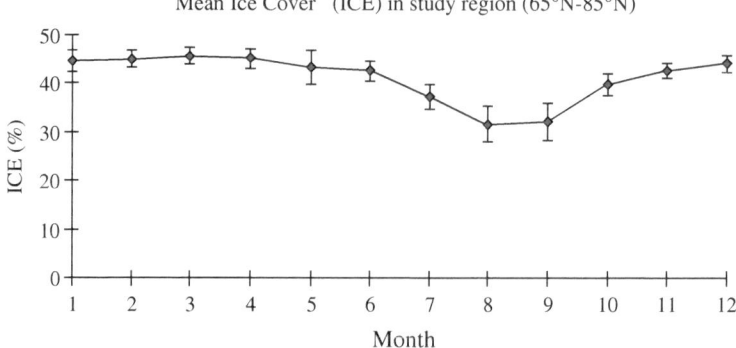

Fig. 2.6   Monthly mean ICE for 10 years in the study region

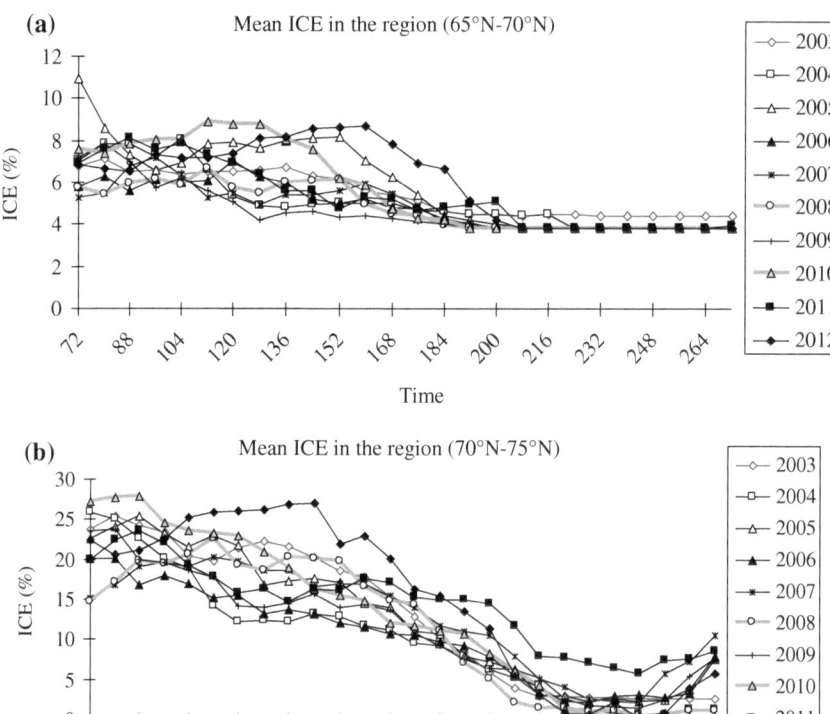

Fig. 2.7   Weekly mean Ice Cover time series in years 2003–2012

second CHL peak. The time span between the two peaks was one more month. Year 2011 had its highest CHL peak in middle of May, and it was not on the MI period, but happened just when the MI stopped decreasing. More MI happened in late March and early April in year 2011. This could be partly the cause of peak of

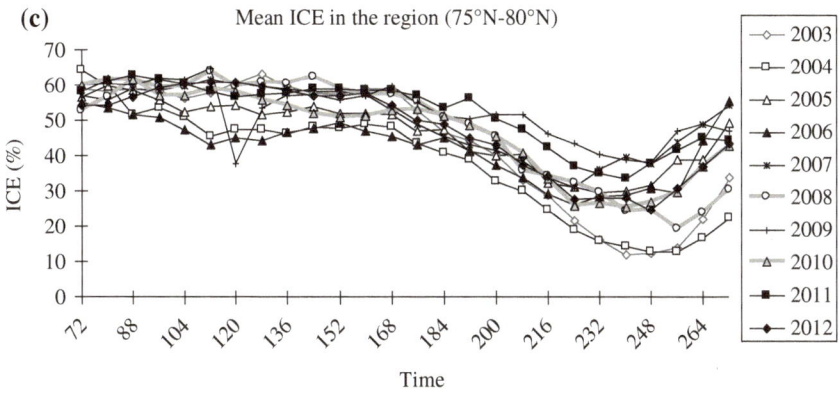

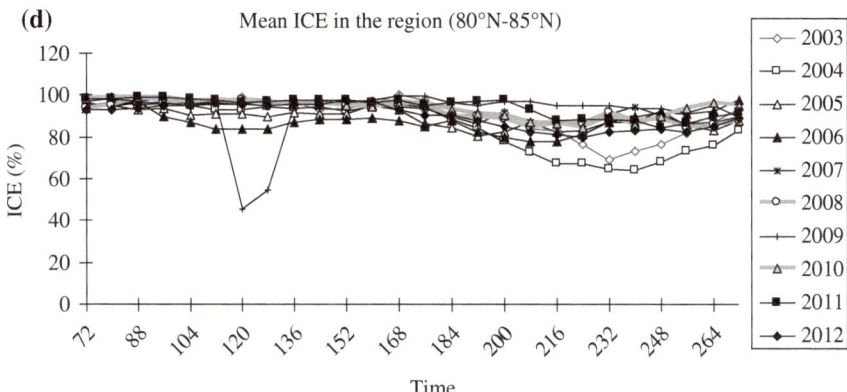

Fig. 2.7 (continued)

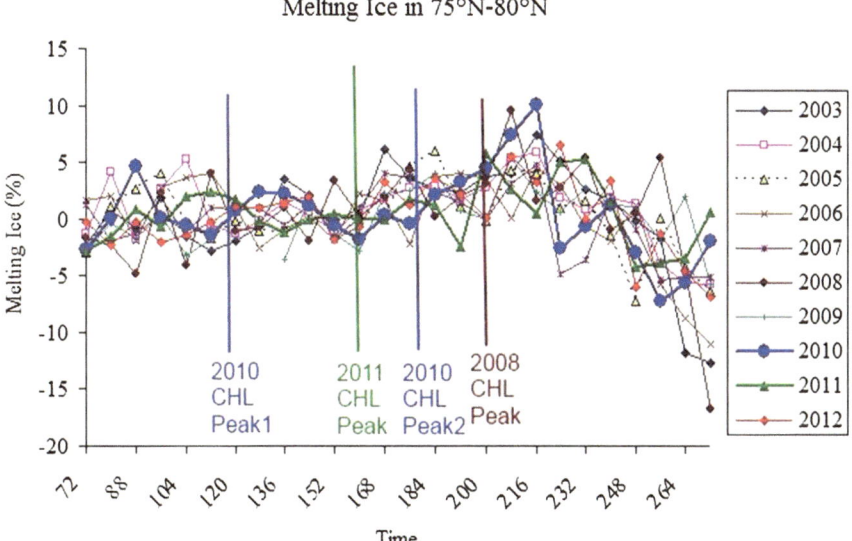

Fig. 2.8 MI profiles for 75°N–80°N in the research area for years 2003–2012

CHL in 2011 in middle of May. However, lower PAR and positive NAO could be the other reasons causing the peak of CHL in 2011. Year 2008 had late CHL peak in early July. The relative low SST could be the reason for the late peak of 2008.

2.3.4 SST, PAR, ICE, and Wind in the 75°N–80°N Region

In our study region, SST had quite strong positive relationship with PAR (Fig. 2.9 for region 75°N–80°N) with PAR 2 months ahead of SST. However, year 2009 had strong negative relationship between PAR and SST.

Looking at 10 years SST profiles in the 75°N–80°N region (Fig. 2.10), the temperature was low in March and gradually increased until July reached to its peak and then started to drop. Year 2003 had the lowest SST during spring and summer. Year 2010 had relative mild SST, while year 2012 had relative higher SST during ice-melting season. Generally, there is an inverse relationship between phytoplankton

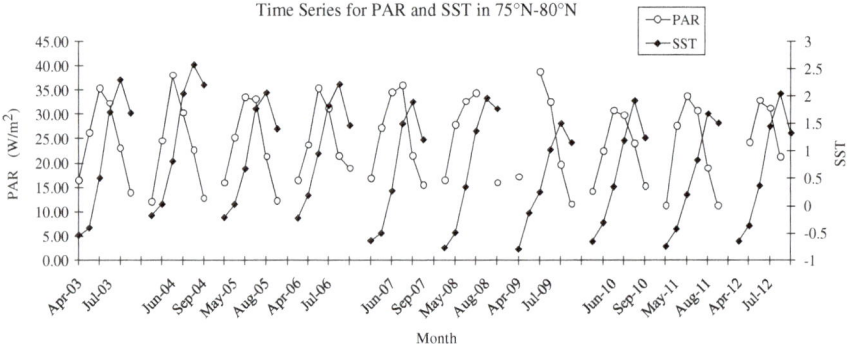

Fig. 2.9   Monthly time series of PAR and SST for the years 2003–2012 in 75°N–80°N

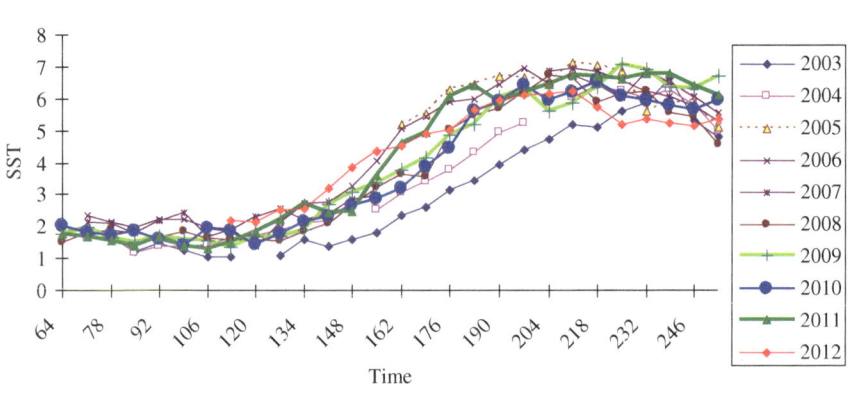

Fig. 2.10   Weekly mean SST in 75°N–80°N for 2003–2012

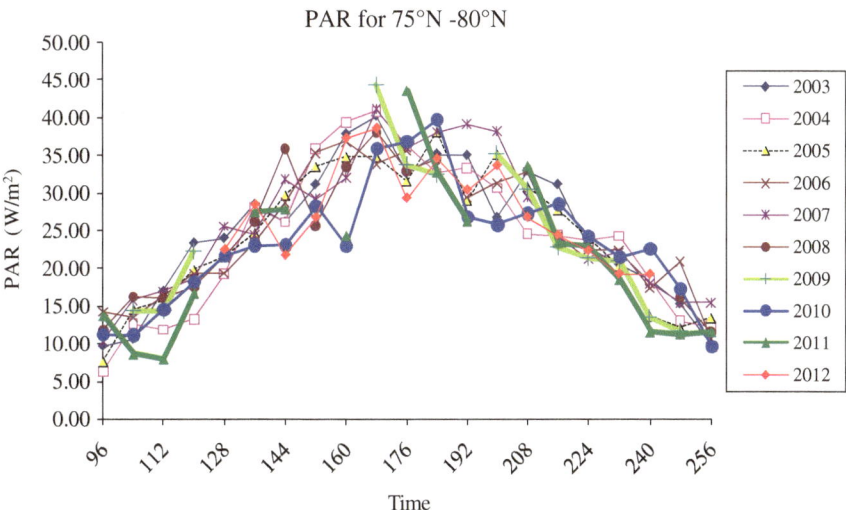

Fig. 2.11 8-day mean PAR in 75°N–80°N for 2003–2012

biomass and SST (Jutla et al. 2009). PAR inter-annual profile for the 10 years in the study region is shown in Fig. 2.11. Year 2009 had highest PAR in summer (although had some missing values), and Year 2011 had summer peak in June. Year 2010 had a dip in middle May. The lower PAR and relative low SST appeared in middle of May favoured growth of phytoplankton biomass in year 2010.

Wind speed in middle of March in 2010 was much higher than other years (Fig. 2.12a). Wind direction generally was southeast direction (Fig. 2.12b). Wind direction in the early spring of 2010 changed from southeast to southwest direction (Fig. 2.12b). That possibly brought MI water from south to north and brought runoff melting water from the east coast of Greenland up to north (79°N–80°N). Year 2010 had relative higher wind speed, and year 2011 had second higher wind speed in spring. Spring wind direction in the both years changed from southeast to southwest with year 2010 changed earlier, and year 2011 changed later but stayed longer in southwest direction. That could explain the Fig. 2.3 that higher CHL peak came earlier in year 2010 and later in year 2011. With year 2011, CHL peak higher than year 2010 within May and June could be due to the longer period of wind direction from southwest.

Yearly MI is calculated in the study region (Fig. 2.13). The positive value shows the MI, and the negative value shows the ice was increasing. The hollow dot line is purely total MI for the year (ignoring the ice increasing amount). The solid dot line includes the MI (positive value) and increasing ice (negative value). Year 2004 had the largest MI through the year. The second largest MI is year 2003. Year 2009 had the least MI and more increased ice. Year 2010 had relative more MI in recent years. In recent 3 years, year 2010 had more MI and 2011 through 2012 had less MI. Table 2.3 lists the weekly ICE trends in different region for the 10 years (March to September). The increasing rate of ICE is insignificant.

Figure 2.14 is the 10 years MI and ice cover for the first half year from day 72–160 (middle of March to middle of May) in the study region. The hollow dot

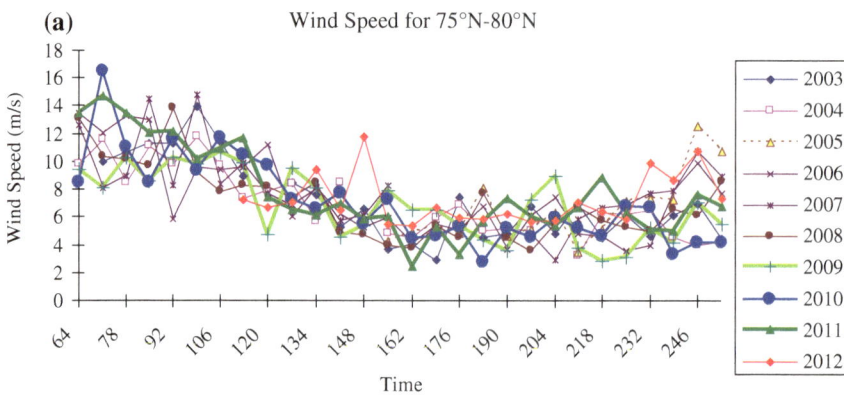

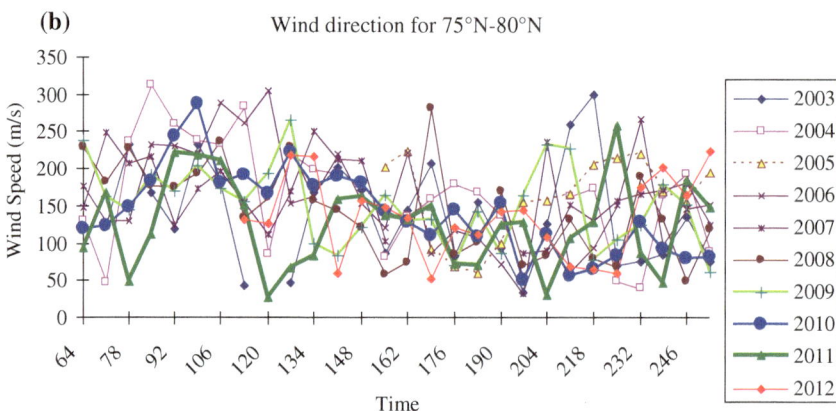

Fig. 2.12 Wind speed and direction in 75°N–80°N for years 2003–2012

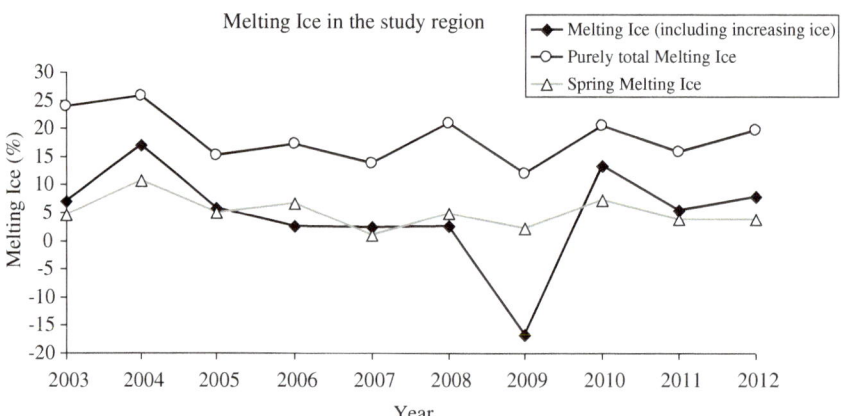

Fig. 2.13 Yearly MI amount for years 2003–2012

Table 2.3 The regression equations of mean ice cover for different subregions

1	65°N–70°N	$y = 0.00002x + 4.4928$
2	70°N–75°N	$y = 0.0005x - 5.7024$
3	75°N–80°N	$y = 0.0022x - 38.388$
4	80°N–85°N	$y = 0.0015x + 31.066$

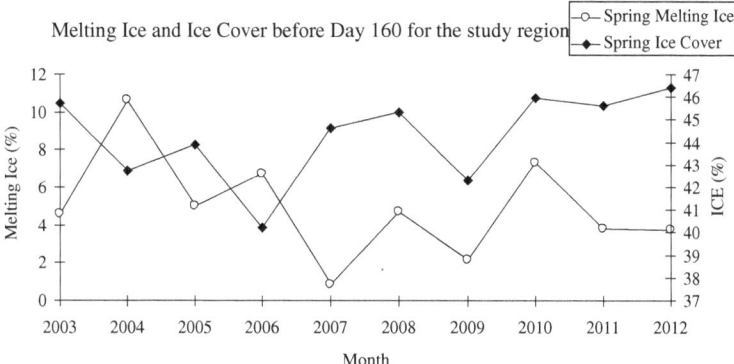

Fig. 2.14 Yearly MI and ICE before day 160 for the study region

line is purely total MI in spring. The solid dot line is the ice cover (ICE) for the same period. Year 2004 had the largest MI, and Year 2010 had the second largest MI. Years 2003, 2010, through 2012 had more increased ICE. That explained year 2010 had higher and early CHL peak. The more ice algae would cause higher CHL concentration in 2010. The general trend of MI in the study region was decreasing in the last 10 years, while ICE increased in general for the spring and early summer.

2.3.5 The Correlation and Regression Analysis Between CHL and ICE

CHL actually increased 1.75 % from 2003 to 2012 in 75°N–80°N region. If only consider the spring and early summer (March to middle of May, up to day 168), the tendency line of CHL is: CHL = 0.1078x, where x is time (see Fig. 2.15). That means in the last 10 years in the region 75°N–80°N, CHL increased 10 % during spring and early summer.

Figure 2.16 is the 10 years mean monthly CHL and ICE in 75°N–80°N region. The general trends of CHL and ICE are as follows: with the decreasing of ICE, CHL was increasing during spring and early summer and reached to its peak in May (for years 2003 and 2010) and in June (for years 2004, 2005, 2006, 2007, and 2011) and in July (for years 2008, 2009, and 2012). In the first 2 months, MI contributed the algae ice to the production of CHL, but after 2 months, the MI water usually did not have significant contribution to the CHL production. Ice did

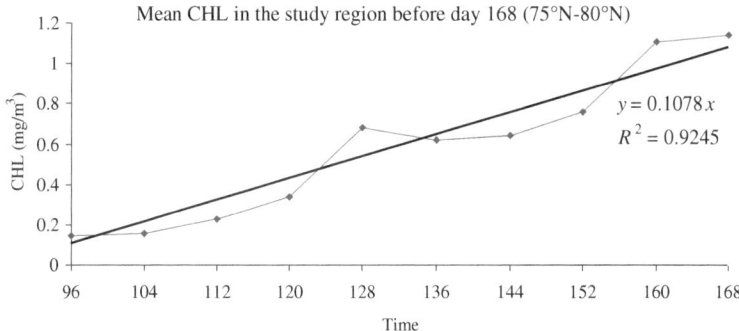

Fig. 2.15 Mean CHL and the tendency line in 75°N–80°N before day 168

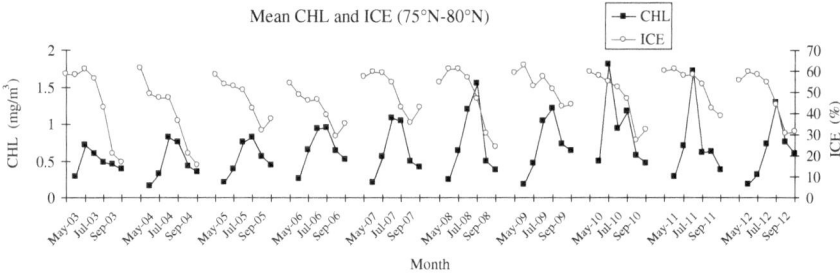

Fig. 2.16 Monthly mean CHL and ICE in the region 75°N–80°N

not melt until May in years 2007, 2008, 2009, and 2012. However, year 2010 had consistent MI from April to August, and CHL was higher in spring in 2010 for monthly data.

Table 2.4 is the correlation coefficient for the CHL and ICE for the years 2003–2012. Generally, they had negative correlations in year 2009 and much significant than other years.

The peak of CHL was about 2 months behind of ICE in 65°N–70°N region, and CHL was about 3 months behind of ICE in other northern regions. Generally, with ice decreasing, CHL would increase from April to its peak in June or July (years 2006, 2008, 2009, and 2012). Year 2010 was quite special with CHL reached to its peak one month earlier in May.

If we shifted ICE 3 months back and aligned with the peak of CHL, there would be quite strong positive coefficient (from 0.51 to 0.68). The correlations of CHL and ICE before and after shifting are shown in Table 2.5.

Table 2.4 The correlation coefficient for CHL and ICE in the 10 years

2003	2004	2005	2006	2007	2008	2009	2010	2011	2012
0.03	−0.21	−0.20	−0.33	0.11	0.07	−0.53	−0.10	0.05	−0.43

Table 2.5 The correlation coefficient of CHL and ICE before and after shifting (20°W–10°E)

	65°N–85°N	65°N–70°N	70°N–75°N	75°N–80°N	80°N–85°N
Before	−0.08	−0.13	−0.11	0.13	−0.43
After	0.68	0.74	0.51	0.60	0.56

Table 2.6 The regression analysis for CHL and ICE(75°N–80°N)

Variable	Coefficient	Std. error	t-statistic	P value
Constant (C)	−0.0447	0.1659	−0.2692	0.7888
ICE	0.0145	0.0033	4.3668	0.0001

Dependent variable: CHL
$R^2 = 0.254$
F-statistic: 19.069

After shifting the CHL 3 months back, the regression analysis and F-statistic checking in 75°N–80°N are shown in Table 2.6.

The correlation analysis for CHL and MI is shown in Fig. 2.17. Different from ICE and CHL had 3 months time lag, MI and CHL had no time lag and their correlation coefficient is 0.4. With MI increasing, the CHL would increase up to its peak. Year 2010 was different. The CHL reached to its peak 2 months before MI. The second peak of MI reached to the same time with the CHL peak. CHL in year 2011 reached to its peak 1 month ahead of MI peak.

Eviews statistics software (Pang 2007) is used to do regression analysis between CHL and ICE.

The regression equation for CHL and ICE is as follows:

$$CHL = -0.045 + 0.0145ICE \ (75°N–80°N) \tag{2.1}$$

The goodness of fit $R^2 = 0.254$ is not a very good fit. Under given significance level $\alpha = 0.05$, t value rejects the hypothesis. The P value in the Table 2.6 shows very good significance. By inspection, we found out that F value is 19.07 > 4.00 (critical

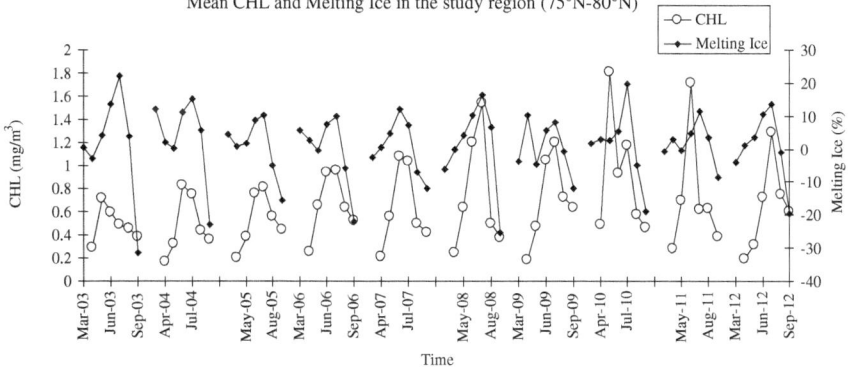

Fig. 2.17 Monthly mean time series for CHL and MI in 75°N–80°N

Table 2.7 Regression analysis for CHL and ICE after shifting

	Regression equation	F value	μ
65°N–70°N	CHL = −0.22 + 0.16ICE	94.58	0.59
70°N–75°N	CHL = 0.12 + 0.03ICE	30.71	0.32
75°N–80°N	CHL = −0.24 + 0.01ICE	5.99	0.13

value, not shown in the table). That means the regression equation is significant. The regression analysis for CHL and ICE in other regions is shown in Table 2.7.

The goodness of fit R^2 is better in region 65°N–70°N and worse in northern region. We use EViews set up a distributed lag model.

Table 2.8 shows the ICE(-3) has lowest P value (0.0003) for CHL regression coefficient test. That means when ICE lagged 3 months behind, ICE had the most significant influence on CHL. This is consistent with the previews results.

The unit root-test is in Table 2.9.

Unit root-test for CHL (Table 2.8) shows that under 1, 5, and 10 % three significant levels, the Mackinnon critical values of unit root-test are −3.5402, −2.9092, and −2.5922, respectively. The t test statistical value (−1.5844) is greater than the critical values; hence, we cannot refuse original hypothesis. This shows CHL had unit root which was non-stationary sequence. We then do unit root-test for the first-order differential sequence.

Table 2.10 shows that t test statistical value −8.8268 is less than all critical values; hence, it can refuse original hypothesis. This shows first-order differential sequence of CHL has no unit root, and it is stationary sequence. Hence, CHL sequence is an integrated of order.

Table 2.8 Distributed lag regression analysis result for CHL and ICE (20°W–10°E)

Variable	Coefficient	Std. error	t-statistic	P value
C	−0.8824	0.2520	−3.5011	0.0009
ICE(−1)	0.01035	0.0035	2.9376	0.0047
ICE(−2)	0.01065	0.0041	2.5925	0.0119
ICE(−3)	0.00159	0.0041	3.8404	0.0003
ICE(−4)	−0.0062	0.0035	−1.7706	0.0816

Dependent variable: CHL
$R^2 = 0.59$
F-statistic: 21.9454
Prob (F-statistic): 0.0000

Table 2.9 Unit root-test for CHL (Null hypothesis: CHL has a unit root)

		t-statistic	Prob.
Augmented Dicky-Full test statistic		−1.5844	0.4844
Test critical values	1 % level	−3.5402	
	5 % level	−2.9092	
	10 % level	−2.5922	

Table 2.10 First-order differential sequence D(CHL) unit root-test for CHL (Null hypothesis: D(CHL) has a unit root)

		t-statistic	Prob.
Augmented Dicky-Full test statistic		−8.8268	0.0000
Test critical values	1 % level	−3.5402	
	5 % level	−2.9092	
	10 % level	−2.5922	

Table 2.11 First-order differential sequence D(ICE) unit root-test for ICE

		t-statistic	Prob.
Augmented Dicky-Full test statistic		−3.9979	0.0026
Test critical values	1 % level	−3.5402	
	5 % level	−2.9092	
	10 % level	−2.5922	

Null hypothesis: D(ICE) has a unit root

Table 2.12 Regression residuals sequence test for CHL and ICE

		t-statistic	Prob.
Augmented Dicky-Full test statistic		−5.2406	0.0000
Test critical values	1 % level	−2.5989	
	5 % level	−1.9456	
	10 % level	−1.6137	

Using the same method, we also found first-order differential sequence of ICE D(ICE) is also a stationary sequence (Table 2.11). Hence, ICE sequence is an integrated of order.

Next is finding if CHL and ICE had co-integration relationship? We do the regression analysis for the two variables: CHL and ICE, then check the smoothness of the regression residuals.

Table 2.12 shows that t test statistical value is −5.2406, less than the correspondent critical value. It shows the residuals sequence does not have unit root, it is stationary sequence. That is, CHL and ICE had co-integration relationship, that means CHL and ICE had long-term equilibrium relationship.

2.3.6 The Correlation Analysis Between NAO and CHL

Monthly NAO from years 2003 to 2012 is shown in Fig. 2.18.

NAO had inter-annual variations. Apart from year 2010 where NAO was negative throughout the year, NAO in other years had more fluctuations through one year. The negative NAO indicated the cold air in European and milder in Greenland. The mild Greenland air would lead more MI from the east coast of Greenland to Greenland Sea. That explains more MI happened in year 2010, and hence, higher and earlier CHL blooms occurred.

Figure 2.19 is the 20 years' time series of CHL and NAO in region 75°N–80°N. In general, CHL had negative relationship with NAO. The correlation coefficient for the 10 years is −0.43865.

We still use Eviews to do regression analysis for CHL and NAO.

Table 2.13 is the regression analysis for CHL and NAO in 75°N–80°N region (Table 2.14).

The regression equation between CHL and NAO is:

$$CHL = 0.19 - 0.16NAO\ (75°N–80°N) \qquad (2.2)$$

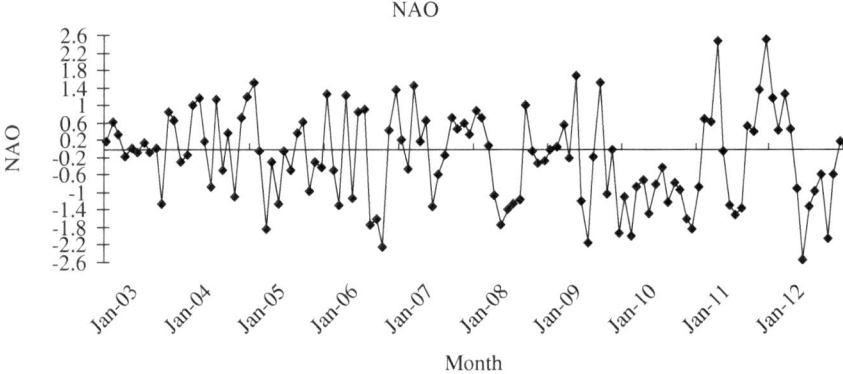

Fig. 2.18 NAO monthly mean for years 2003–2012

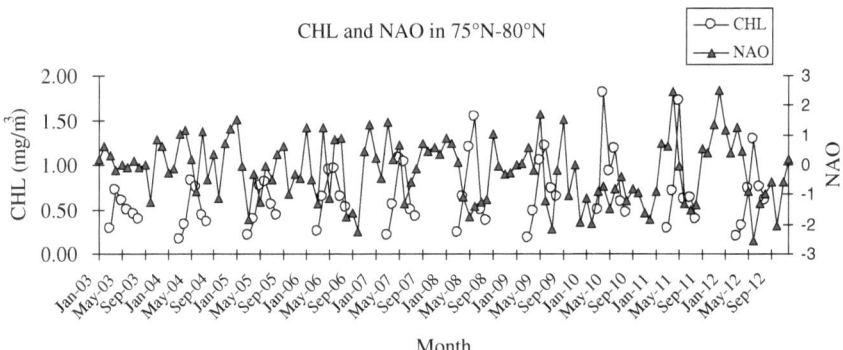

Fig. 2.19 Monthly mean CHL and NAO in years 2003–2012 in 75°N–80°N

Table 2.13 Regression analysis for CHL and NAO in different sub-region

	Regression equation	F value	μ
65°N–70°N	CHL = 0.60–0.07NAO	4.64	0.06
70°N–75°N	CHL = 0.52–0.10NAO	3.70	0.05
80°N–85°N	CHL = 0.39–0.01NAO	0.18	0.004

Table 2.14   Regression analysis for CHL and NAO (75°N–80°N)

Variable	Coefficient	Std. error	t-statistic	P value
Constant (C)	0.5965	0.0456	13.0827	0.0000
NAO	−0.158	0.0425	−3.7174	0.0005
$R^2 = 0.1924$			F-statistic: 13.8189	

Dependent variable: CHL

Giving significant level $\alpha = 0.05$, after t test and F test, P value is smaller than α (P value $= 0.0005$). Hence, the regression equation is significant. Other regression equation is listed in Table 2.13. The small value of R^2 could be due to the different cycle of the two parameters.

2.3.7 Correlations of MI and NAO

It is found MI had better correlation with NAO rather than ICE and NAO. In region 75°N–80°N, the 10 years monthly MI and NAO time series is shown in Fig. 2.20. There was a positive correlation relationship in some time period, although it was not consistently always positive. It is noticed that NAO was 3 months ahead of MI. This result is confirmed by Eviews. Figure 2.21 is the weekly two time series for year 2010 in the same region. The detailed time series shows there was obvious correlation between the two (Fig. 2.21a). If we shift MI 3 weeks ahead, the MI and NAO were negatively correlated (Fig. 2.21b). The correlation coefficient is −0.57. It means with the increase of NAO, MI would decrease. On the other hand, with decrease of NAO, MI would increase.

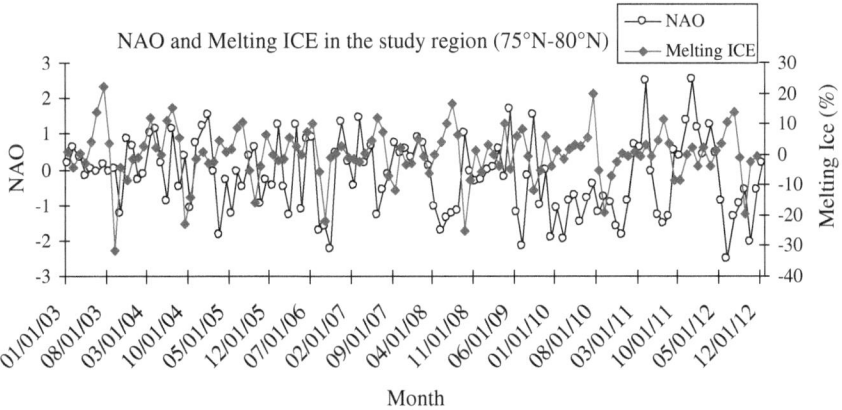

Fig. 2.20   10 years NAO and MI time series in 75°N–80°N

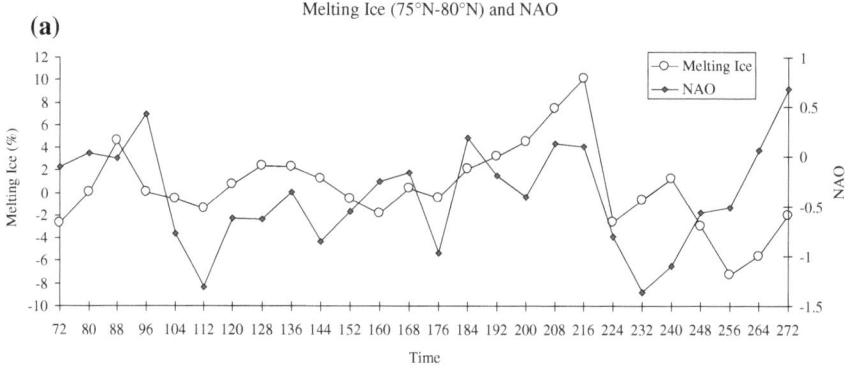

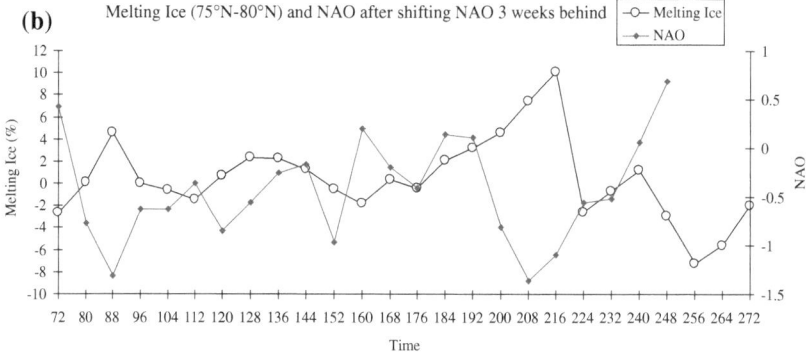

Fig. 2.21 Weekly time series of MI and NAO in 75°N–80°N in year 2010

If considering the positive correlations for monthly data, the time lag is much larger. For years 2011 and 2012, the time lag between MI and NAO is 3 months. After shifting MI 3 months ahead, the high positive correlation is shown in the two figures (Fig. 2.22). The correlation coefficients for these two years are 0.81 and 0.84, respectively, for 2011 and 2012.

2.3.8 The Correlation Analysis Among CHL, MI, and NAO

We still study the correlation among CHL, NAO, and MI in 75°N–80°N region (Table 2.15).

We have the regression equation:

$$CHL = 0.054 - 0.111NAO - 0.012ICE(-3)(75°N–80°N) \qquad (2.3)$$

Table 2.16 confirmed ice and NAO had 3 months' time lag with NAO 3 months ahead of MI. The regression equation is significant. That means NAO and MI

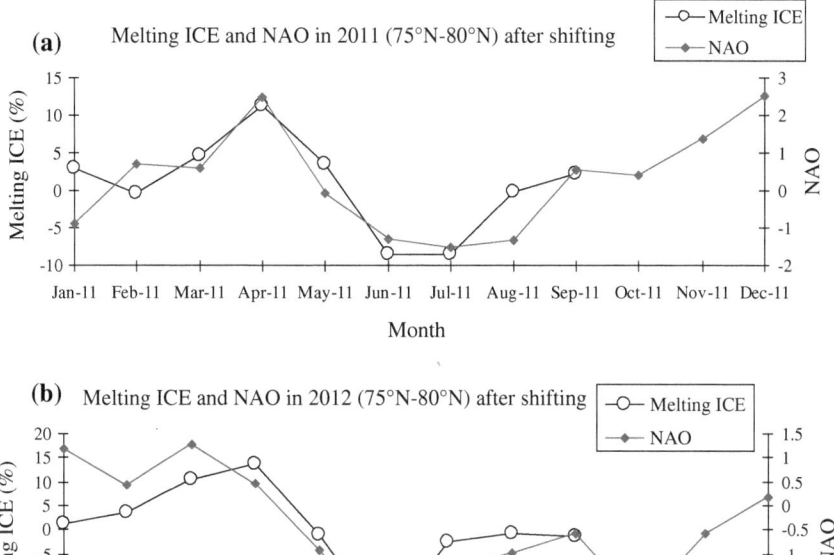

Fig. 2.22 Month mean MI and NAO after shifting in 75°N–80°N for years 2011 and 2012

Table 2.15 Regression analysis for CHL, NAO, and MI.\(ICE) (Dependent variable: CHL)

Variable	Coefficient	Std. error	t-statistic	P value
Constant (C)	−0.0544	0.1659	0.3368	0.7375
ICE(−3)	0.0116	0.0033	3.5066	0.0009
NAO	−0.1118	0.0412	−2.7092	0.0090
$R^2 = 0.3419$			F-statistic: 14.2838	

Table 2.16 Regression analysis for CHL, NAO, and melted ice in different subregions

	Regression equation	F value	μ
65°N–70°N	CHL = −0.227 + 0.005NAO + 0.158ICE(−2)	46.63	0.59
70°N–75°N	CHL = 0.266 − 0.054NAO + 0.021ICE(−3)	6.31	0.16
80°N–85°N	CHL = −0.177 − 0.035NAO + 0.006ICE(−3)	5.69	0.22

had significant influence on CHL. The southern region had the more correlated correlations.

2.4 Conclusions

The distributions and correlation analysis between CHL and ICE, CHL and MI, NAO and ICE, NAO and MI, and CHL and NAO are all studied. The MI played a significant role on promoting the growth of CHL. We are more focused on the northern region (75°N–80°N), where ice melted more. It was unusual to find CHL was higher near 80°N in the study region due to the enhanced water column stability and more MI, less salinity in this region. The peaks of CHL in 2010 happened very early and much higher than other year (in the same time). We tried to find the reason for that. The first high peak in early spring could be due to the higher wind speed. Spring wind direction in the year changed from southeast to southwest direction brought more MI water in the northern region, hence promoting the CHL concentrations. MI in year 2010 was much more than other recent years. Moreover, when temperature was warm in middle of May, the relative mild SST and lower PAR profile in year 2010 favoured the growth of phytoplankton biomass.

The peak of CHL was about 2 months behind of ICE in 65°N–70°N region, and 3 months behind of ICE in other northern regions. After shifting ICE 3 months back, the correlation between CHL and ICE was 0.68. That means ICE influenced CHL and had positive correlations with CHL.

NAO had almost negative index in year 2010, and it refers the mild Greenland air would lead to more MI in that year. NAO had negative correlations with CHL; with lower NAO, stronger CHL would appear. MI had better correlations with NAO than Ice cover with NAO. In year 2010, MI and NAO were negatively correlated with NAO 3 weeks ahead of MI. If shifted NAO 3 months behind, there would be higher correlation coefficients (0.81 and 0.84) between NAO and MI in years 2011 and 2012, respectively.

We focus the region in 75°N–80°N for correlation analysis and found out that NAO and MI had significant influence on CHL.

References

Cherkasheva, A., Nöthig, E. M., Bauerfeind, E., Melsheimer, C., & Bracher, A. (2014). From the chlorophyll-a in the surface layer to its vertical profile: a Greenland Sea relationship for satellite applications. *Ocean Science, 9*, 431–445.

Cui, S., He, J., He, P., Zhang, F., Lin, L., & Ma, Y. (2012). The adaptation of Arctic phytoplankton to low light and salinity in Kongsfjorden (Spitsbergen). *Advances in Polar Science, 23*, 19–24.

Ericken, H., Ackley, S. F., Richter-Menge, J. A., & Lange, M. A. (1991). Is the strength of sea-ice related to its chlorophyll content? *Polar Biology, 11*, 347–350.

Gradinger, R. R., & Baumann, M. E. M. (1991). Distribution of phytoplankton communities in relationship to the large scale hydrographical regime in Fram Strait 1984. *Marine Biology, 111*, 311–321.

Horner, R., Ackley, S. F., Dieckmann, G. S., Gulliksen, B., Hoshiai, T., Legendre, L., et al. (1992). Ecology of sea-ice biota. I. Habitat, terminology, and methodology. *Polar Biology, 12*, 417–427.

Jassby, A. D., & Platt, M. E. (1976). Mathematical formulation of the relationship between photosynthesis and light for phytoplankton. *Limnology and Oceanography, 21*, 540–547.

Jutla AS, Akanda AS, and Islam S (2009) Relationship between Phytoplankton, Sea Surface Temperature and River Discharge in Bay of Bengal. General Assembly of the European Geosciences Union, Vienna, Austria, April 19-24.

Lara, R. J., Lara, R. J., KATTNER, G., Tillmann, U., & Hirche, H. J. (1994). The North East Water polynya (Greenland Sea) II. Mechanisms of nutrient supply and influence on phytoplankton distribution. *Polar Biology, 14*, 483–490.

Leu, E., Soreide, J. E., Hessen, D. O., Falk-Petersen, S., & Berge, J. (2011). Consequences of changing sea-ice cover for primary and secondary producers in the European Arctic shelf seas: Timing, quantity, and quality. *Progress in Oceanography, 90*, 18–32.

Matrai, P. A., & Vernet, M. (1997). Dynamics of the vernal bloom in the marginal ice zone of the Barents Sea: Dimethyl sulfide and dimethylsulfoniopropionate budgets. *Journal of Geophysical Research: Ocean, 102*, 22965–22979.

Olli, K., Riser, C. W., Wassmann, P., Ratkova, T., Arashkevich, E., & Pasternak, A. (2002). Seasonal variation in vertical flux of biogenic matter in the marginal ice zone and the central Barents Sea. *Journal of Marine Systems, 38*, 189–204.

Pabi, S., van Dijken, G. L., & Arrigo, K. R. (2008). Primary production in the Arctic Ocean, 1998–2006. *Journal of Geophysical Research: Oceans, 113*, C08005. doi:10.1029/200 7JC004578.

Pang, H. (2007). *Econometrics* (pp. 265–284). Beijing: Science Publishing Press.

Parsons, T. R., Maita, Y., & Lalli, C. M. (1984). *A manual of chemical and biological methods for seawater analysis*. Oxford and New York: Pergamon Press.

Qu, B., Gabric, A. J., Lu, H., & Lin, D. (2014). Spike in phytoplankton biomass in Greenland Sea during 2009 and the correlations among chlorophyll-a, aerosol optical depth and ice cover. *Chinese Journal of Oceanology and Limnology, 32*(2), 241–254.

Qu, B., Gabric, A. J., & Matrai, P. A. (2006). The Satellite-Derived Distribution of Chlorophyll-a and its Relation to Ice Cover, Radiation and Sea Surface Temperature in the Barents Sea. *Polar Biology, 29*, 196–210.

Schneider, W., & Budéus, G. (1994). The North East Water Polynya (Greenland Sea). I. A physical concept of its generation. *Polar Biology, 14*, 1–9.

Serreze, M., Walsh, J., Chapin, F., Osterkamp, T., Dyurgerov, M., Romanovsky, V., et al. (2000). Observational evidence of recent changes in the northern high latitude environment. *Climate Change, 46*, 159–207.

Soreide, J. E., Leu, E., Graeve, M., & Falk-Petersen, S. (2010). Timing of blooms, algal food quality and Calanus glacialis reproduction and growth in a changing Arctic. *Global Change Biology, 16*, 3154–3163.

Vancoppenolle, M., Bopp, L., Madec, G., Dunne, J., Ilyina, T., Halloran, P. R., et al. (2013). Future Arctic Ocean primary productivity from CMIP5 simulations: Uncertain outcome, but consistent mechanisms. *Global Biogeochemical Cycles, 27*(3), 605–619.

Wassmann, P., Ratkova, T. N., Andreassen, I., Vernet, M., Pedersen, G., & Rey, F. (1999). Spring Bloom Development in the Marginal Ice Zone and the Central Barents Sea. *Marine Ecology, 20*, 321–346.

Chapter 3
Aerosol Optical Depth, Ice Cover, and Cloud Cover

Abstract This chapter investigated the relationships between aerosol optical depth, sea-ice cover (ICE), and cloud cover (CLD) in the Greenland Sea in 20°W–10°E, 65°N–85°N during the period 2003–2012. We focused more on 70°N–80°N and divided it into two 5° zonal apart. Remote sense satellite data were used to do correlation analysis. Enhanced statistics methods are used for correlation and regression analysis. According to the 10-year data, AOD was high in spring, and low in summer, and it rose back again in autumn. AOD content was generally higher in southern region (70°N–75°N) than the northern region (75°N–80°N). AOD and ICE had positive correlations, while AOD and CLD had negative correlations. The peaks of ICE and CLD were all 1 month earlier than the peak of AOD. That indicates both ice cover and cloud cover all had influenced on AOD content. After shifting ICE and CLD 1 month later, they both had long-term equilibrium relationship with AOD. The correlation between AOD and ICE was stronger than the correlation between AOD and CLD, indicating that the aerosols in Arctic mostly came from the sea ice rather than from the air cloud. Melting ice (MI) resulted in the increasing of the AOD content.

Keywords Aerosol optical depth (AOD) · Cloud cover (CLD) · Ice cover (ICE) · Melting ice (MI) · Coupling · Peak

3.1 Introduction

3.1.1 Aerosol and Cloud

An aerosol is a suspension of solid or liquid particles in the air. Sulfate, nitrate, organics, soil dust, sea salt, ammonia, black carbon, and trace metals are all aerosols. Those include natural gases and anthropogenic gases. Nearly 50 % of fine aerosols are from anthropogenic sources (such as fossil fuel burning, dust from fires smoking, domestic fuels, and bagasse burning). There are about more than 50 % off the fine aerosols is sulfate (Ramanathan et al. 2007). Large aerosols such

B. Qu, *The Impact of Melting Ice on the Ecosystems in Greenland Sea*, SpringerBriefs in Environmental Science, DOI 10.1007/978-3-642-54498-9_3 as solid grains and sea salt are too heavy to lift up in the air, so they have no much effect on the atmosphere. Cloud droplets are the aerosols with its radius roughly 10–20 microns. It could suspend up the air, and by condensation of vapor to its relative humidity reduced to near 100 % (Ramanathan et al. 2007), cloud condensation nuclei (CCN) process is formed.

Aerosols are important to climate. They scatter and absorb radiation in the atmosphere; hence, they change the microphysical structure and cloud lifetimes as well. The scattering of solar radiation acts to cool the earth, while absorption of solar radiation acts to warm the air directly. Clouds act both scattering solar radiation and absorbing thermal radiation. Aerosol optical depth is a measure of the strength of interaction of clouds with radiation. Low-altitude warm liquid clouds mainly scatter solar radiation and cool the planet, while the high-altitude ice cloud mainly absorb thermal radiation and warm the planet (Ramanathan et al. 2007). Hence, changes in aerosol concentrations would greatly influence the radiation balance and the climate.

It was reported that the global average change in surface temperature is related to the radiative forcing. Therefore, climate sensitivity is related to the changes in clouds, water vapor, sea-ice cover (ICE), and snow.

The sulfate aerosols mainly offset the climate warming by cooling down the temperature. The most effect area of sulfate aerosol is in northern hemisphere Arctic Ocean. If the greenhouse gas emissions remain constant, with the increase of sulfate aerosol in Arctic Ocean, the greenhouse gas impacts would decrease in the future and climate would cool down. If the short lifetime sulfate aerosol did keep pace with the long life greenhouse gases, the global warming would still accelerate. However, with control of greenhouse gases emission, the cooling effect from sulfate aerosol cannot be ignored.

3.1.2 Sea Ice, AOD, Cloudiness, and Radiative Balance

Sea ice is a source of sea salt aerosols, and the rapid decrease of sea ice extend in Arctic Ocean could result in the sea salt aerosol emission increasing, in turn could lead to increase the natural AOD about 23 % (Struthers et al. 2011). Serreze et al. (2007) pointed out that the Arctic ecosystem and ocean circulations are greatly influenced by the retreating of sea ice due to increase in temperature. This in turn would reduce the surface albedo, increasing the sunlight absorbing by the earth surface, and hence lead to further increase in temperature. This formed a positive feedback system. However, by reducing sea-ice extent, the consequences of changing in sea salt aerosol emission in the Arctic are not quite clear (Quinn et al. 2007, 2008).

The physical drivers of sea salt aerosol are the sea-ice cover, surface wind speed, and sea surface temperature (Nilsson et al. 2007). The ice cover plays a significant role among the three drivers. The particle emission fluxes in sea salt aerosol mode are calculated as follows:

$$\langle \text{flux} \rangle = W(c_2 \times \text{sst}^2 + c_1 \times \text{sst} + c_0) \qquad (3.1)$$

where the whitecap fraction W is calculated as follows:

$$W = 3.84 \times 10^{-4} U_{10}^{3.41} \qquad (3.2)$$

here, U_{10} is 10 m horizontal wind speed (Monahan and Muircheartaigh 1980).

The non-sea salt aerosol includes sulfate, particulate organic matter, mineral dust, and black carbon. Gabric et al. (2005) suggested that the gaseous DMS cycle in Arctic Ocean is depended on the sea-ice extent. Here, DMS is the main sulfate aerosol released from ocean in Arctic. The melting of ice caused ice algae contribute significant rate of phytoplankton biomass in Arctic and hence promote the growth of DMS.

Aerosol radiative effects can be separated into two components: natural and anthropogenic. The natural aerosol includes sea salt and mineral dust. The sea salt AOD is higher in winter, and it is a major natural aerosol in autumn and winter (Struthers et al. 2011). In Arctic Ocean, black carbon and mineral dust aerosols are relative much less considering in total AOD. Sulfate contributes about 0.03 to the total AOD (Struthers et al. 2011).

Winter cloud amount changes would alter AOD signal. The increase of winter time cloud fraction would likely decrease the natural AOD in Arctic (Struthers et al. 2011). The radiative balance in Arctic would be greatly impacted by the changes in surface albedo and sea salt aerosol due to the loss of Arctic sea ice. Natural aerosol forcing is generally negative, but positive for the northern Arctic in summer time (May–July), implies the warm signal in these period. This is due to the scattering aerosol over high albedo surface especially over sea-ice area. This is also due to the relative high solar zenith angle. In spring and early summer in Arctic Ocean, the sea ice has as high as 0.7–0.9 or more albedo due to the snow lying on top of the ice. After melting, the albedo would reduce to 0.4–0.7 in the summer period (Struthers et al. 2011). The open water within the sea ice in Arctic also has high albedo. Struthers et al. (2011) did research in Arctic region, and by increasing sea salt aerosol emission would cause increasing of aerosol forcing, here 50 % would come from ice–albedo effect. Changes in surface albedo and cloud amount would greatly influence on the cloud radiative forcing. It was also found that the first indirect aerosol forcing is approximately 10 times the direct aerosol forcing in Arctic.

The increasing of aerosol forcing would cause decreasing the atmospheric temperatures. The sea-ice spray feedback is a negative feedback to the Arctic climate change. The sea-ice spray feedback is greatly linked to the Arctic sea-ice–albedo feedback mechanism; however, the sea-ice spray feedback is most likely not large enough to counteract the ice–albedo feedback; however, it would reduce it. The aerosol–cloud feedback is particularly important for Arctic region. It is emphasized that the cloudiness and cloud radiative forcing are strongly coupled to Arctic sea-ice cover (Struthers et al. 2011).

Satellite data show that there is a positive correlation between total cloud cover (TCC) and AOD (Sekiguchi et al. 2003). Kaufman et al. (2005) found the global annual mean cloud cover (CLD) would increase 3 % due to anthropogenic aerosol. This effect would be almost balance forcing by doubling in CO_2 concentration (Slingo 1990).

The strength of the aerosol–cloud cover relationship is as follows:

$$b = \Delta \ln \text{TCC}/\Delta \ln \text{AOD} \qquad (3.3)$$

This formula shows the relative change in TCC with relative perturbation in AOD (Feingold et al. 2003).

Data and methods are listed in Chap. 2 (Sect. 2.2). In this chapter, we will find the relationship between AOD, ice cover, and cloud cover in the Greenland Sea and determine the dominant effect to the AOD in the study region.

3.2 Results

3.2.1 The AOD Distributions

Eight-day mean AOD in the whole study region (65–85°N, 20°W–10°E) is shown in Fig. 3.1, with higher AOD in spring and lower in summer and further increased in autumn. Vertical bars are the standard deviations. Yearly mean AOD in the study region in the 10 years (2003–2012) is shown in Fig. 3.2. The vertical bars are the standard deviations for each year. Year 2009 had the highest and more stable AOD, and 2010 had the lowest AOD.

Mean AOD (8-day interval) time series in the study region in each year is shown in Fig. 3.3. Generally, spring AOD was higher than summer AOD. Autumn AOD would further decrease. Figure 3.3a is the distribution of AOD in the region 65°N–85°N. Year 2003 had higher level of AOD in spring and summer. It could be due to the big extensive Russia fires happened in the early half year of 2003

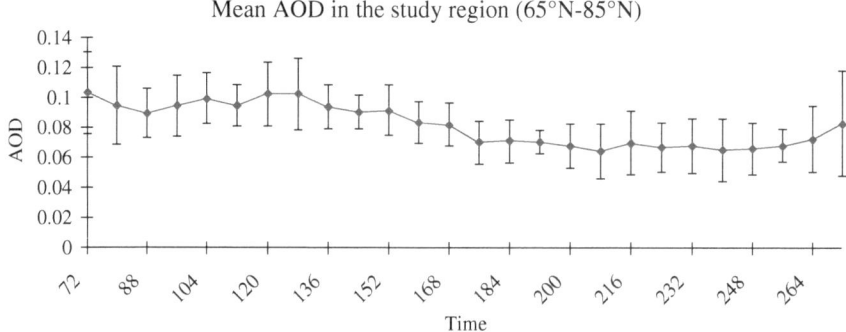

Fig. 3.1 Mean AOD for 8-day interval in the study region averaged for 2003–2012

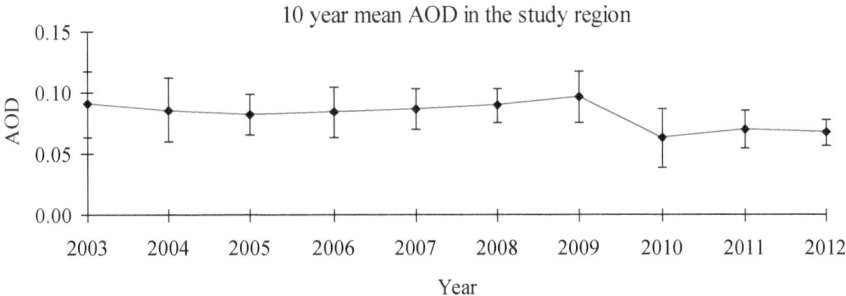

Fig. 3.2   Yearly mean AOD in 2003–2012 in the study region

(Serreze et al. 2000). Year 2011 had a peak in early spring and year 2006 had higher peak of AOD in April. Year 2009 had several high peaks of AOD late summer and autumn. Years 2010 and 2012 had relative lower AOD throughout the year. Year 2008 had relative longer period of higher AOD in spring and autumn.

We focus on the region of 70°N–80°N. The southern region AOD (70°N–75°N) (Fig. 3.3b) was generally higher than northern region (75°N–80°N) (Fig. 3.3c). AOD in March only appeared in southern region, not northern region, due to late start of the sunlight up north. Year 2009 had more significant AOD especially in late summer and early autumn. Year 2003 had second peak in middle of May. Year 2004 had 3 peaks through the year and with the last peak in late summer. Year 2008 had relative high AOD in spring in northern region. Year 2010 had relative low AOD in the whole study region.

Mean AOD along latitude in the study region is shown in Fig. 3.4. Generally, AOD were higher in south and lower in north. Year 2009 had higher AOD within 73°N–80°N. Year 2006 had higher AOD in north of 80°N. Year 2003 had higher AOD in southern region (south of 72°N) especially high in the very south point (65°N–66°N, out of Arctic Ocean). It is obviously that the Russia fire had more impact on the southern region in year 2003. Years 2010 and 2012 had relative lower AOD especially in southern region. Year 2008 had relative higher AOD in 77°N and 71°N. AOD in three-dimensional plots for years 2009 and 2010 in 65–85°N is shown in Fig. 3.5a, b. The higher AOD in late summer and autumn in year 2009 is obvious, while spring and summer high peaks of AOD in 2010 are also shown in the Fig. 3.5b.

3.2.2 Cloud Cover (CLD) Distributions

Mean cloud cover in the study region for the 10 years is shown in Fig. 3.6. Generally, the cloud cover is around 0.8 with increased trend after April up to September. Figure 3.6b is the monthly mean for each year in the region

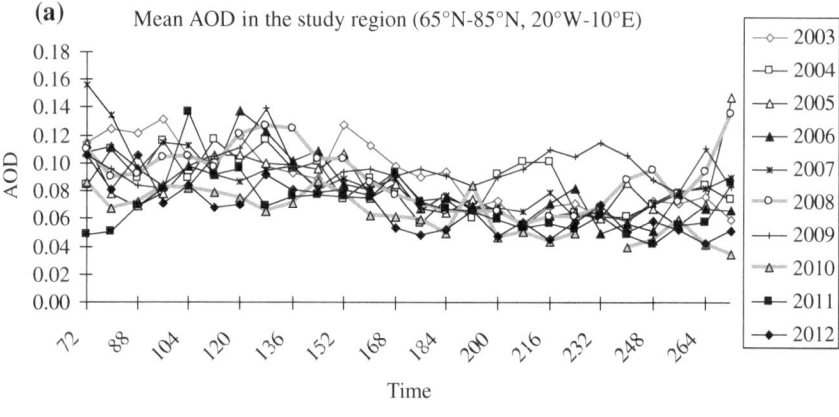

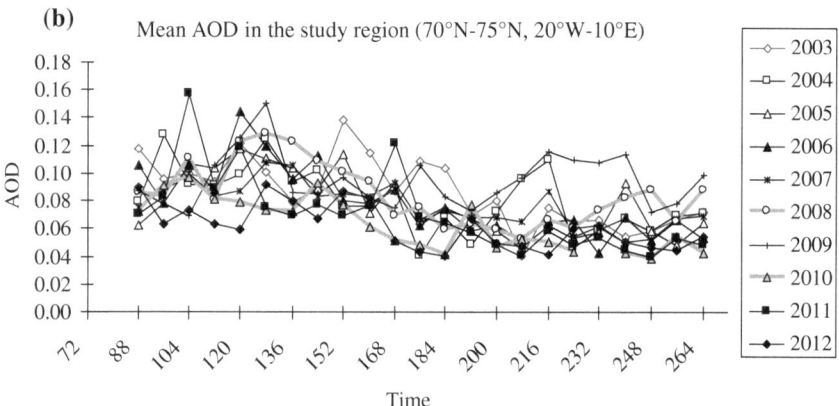

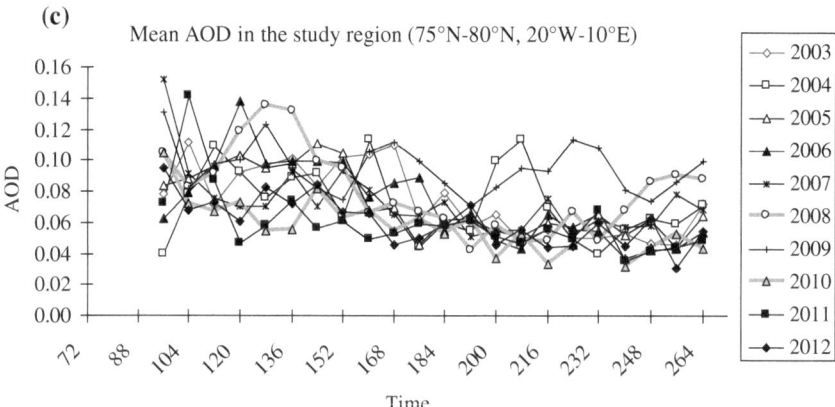

Fig. 3.3 Mean weekly AOD for year 2003–2012 in the region (20°W–10°N): **a** 65°N–85°N; **b** 70°N–75°N; **c** 75°N–80°N

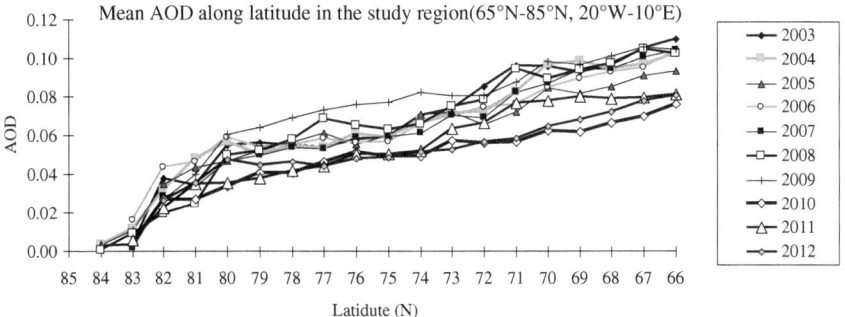

Fig. 3.4 Mean AOD along latitude in the study region

70°N–80°N. Year 2004 had more cloud cover than other years. Year 2009 had the least cloud cover in April. Years 2008 and 2011 had least cloud cover from May to July. April is the month with the most fluctuations, while August and September had the least fluctuations.

3.2.3 The Correlation Analysis Among AOD, ICE, and CLD

3.2.3.1 The Correlation Analysis for AOD and ICE

Monthly mean AOD and ice cover (ICE) for the region 70°N–75°N and 75°N–80°N are showing in Fig. 3.7a, b. Hollow dots represent AOD and solid dots represent ICE. Although there was a time lag between AOD and ICE, they had quite strong positive correlations. The peaks of ICE were generally 1 month ahead of AOD. The ice cover in north was much higher than south (right axis). In 70°N–75°N, ICE was high in year 2004, 2010, and 2012. AOD was lower in years 2010 and 2012, but much higher through the year 2009 (there was three peaks in 2009 and the last peak was in September). In 75°N–80°N, ICE had no much decrease in the recent years apart from year 2006. Year 2010 had much lower AOD and followed the low AOD in years 2011 and 2012. The same as AOD, ICE in year 2009 also had two peaks. AOD in 2008 had the highest peak.

The correlation coefficients between AOD and ICE are listed in Table 3.1 for the 10 years. The correlation coefficient ranged from 0.58 to 0.88 apart from year 2009. Years 2003, 2007, and 2012 had higher correlation coefficient values (0.76–0.88). Year 2009 had negative value −0.43. The early CHL peak (April) and relative late AOD peak (June) in Fig. 3.8a show there were strong negative correlations between AOD and MI in northern region.

The peak times of AOD and ICE are calculated and listed in Table 3.2. Table 3.3 is the lag time table for 75°N–80°N.

Table 3.3 shows the peak time of ICE usually was ahead of AOD for 1 month. We shift AOD 1 month ahead, and the correlation would improve.

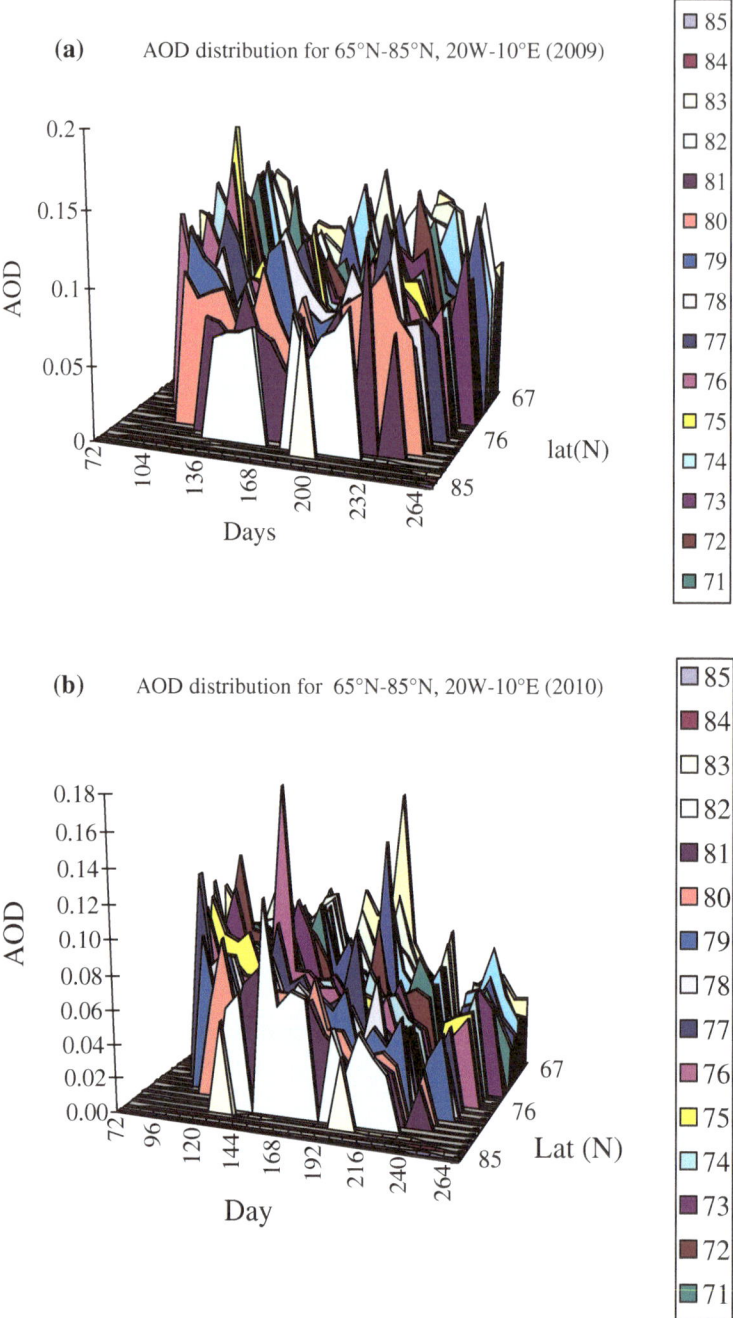

Fig. 3.5 3D AOD for years 2009 and 2010 in the study region. **a** AOD distribution for 65°N–85°N, 20°W–10°E (2009). **b** AOD distribution for 65°N–85°N, 20°W–10°E (2010)

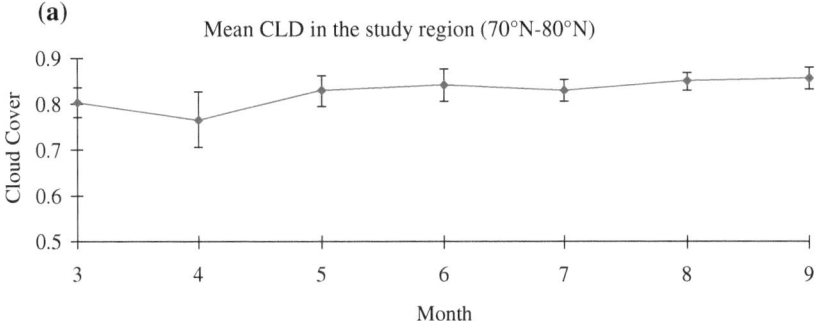

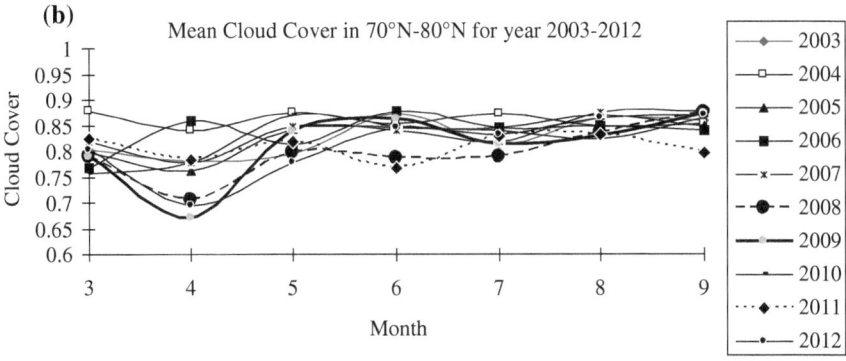

Fig. 3.6 a Monthly mean cloud cover in the study region. The error bars are the standard deviation. **b** Monthly mean cloud cover for each year in the region 70°N–80°N and 20°W–10°E

If we calculate MI by subtracting the ICE data from previous week to this week, AOD and MI would have negative relationship during early spring and summer (Fig. 3.8). When MI increased, the AOD would decrease. When MI reached to peak, AOD reached to its valley. This again confirmed the low AOD in 2010 is caused partly by more MI in the year.

3.2.3.2 The Regression and Lag Analysis for AOD and ICE

The correlation coefficient is calculated using following formula:

$$
P_{xy}(L) = \begin{cases} \dfrac{\sum_{k=0}^{N-|L|-1} (x_{k+|L|} - \bar{x})(y_k - \bar{y})}{\sqrt{\left[\sum_{k=0}^{N-1} (x_k - \bar{x})^2\right]\left[\sum_{k=0}^{N-1} (y_k - \bar{y})^2\right]}}, & \text{when} \quad L < 0 \\[4ex] \dfrac{\sum_{k=0}^{N-L-1} (x_k - \bar{x})(y_{k+L} - \bar{y})}{\sqrt{\left[\sum_{k=0}^{N-1} (x_k - \bar{x})^2\right]\left[\sum_{k=0}^{N-1} (y_k - \bar{y})^2\right]}}, & \text{when} \quad L \geq 0 \end{cases} \tag{3.4}
$$

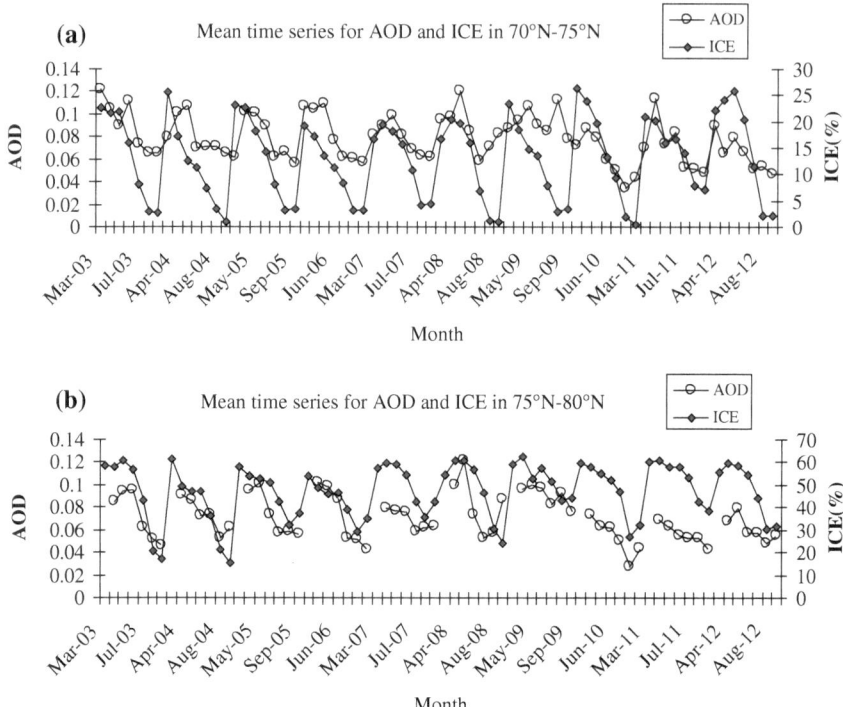

Fig. 3.7 Monthly mean AOD and ice cover time series in 70°N–75°N and 75°N–80°N

Table 3.1 The correlation coefficient for AOD and ICE

2003	2004	2005	2006	2007	2008	2009	2010	2011	2012
0.88	0.71	0.72	0.84	0.78	0.58	−0.43	0.97	0.68	0.76

where $\bar{x}$ and $\bar{y}$ are mean values of ICE and AOD time series.

We are more interested in the 75°N–80°N region where more ice cover and more MI happened in this region. Time lag regression analysis is first carried out using EViews statistical software (Pang 2007) to find out the exact lag time between AOD and ICE.

Different from all other lag, the P value of ICE(-1) is 0.001 which is significantly smaller than 0.05 (Table 3.4). Hence, if ICE lagged 1 month behind, ICE and AOD would have more significant relationship. This is coincident with the previous results.

Table 3.5 shows that P value of ICE(−1) is 0.0000 which is significant smaller than 0.05; hence, there is a significant correlation between AOD and ICE(−1). The regression equation is as follows:

$$AOD = -0.011532 + 0.001530\,ICE \tag{3.5}$$

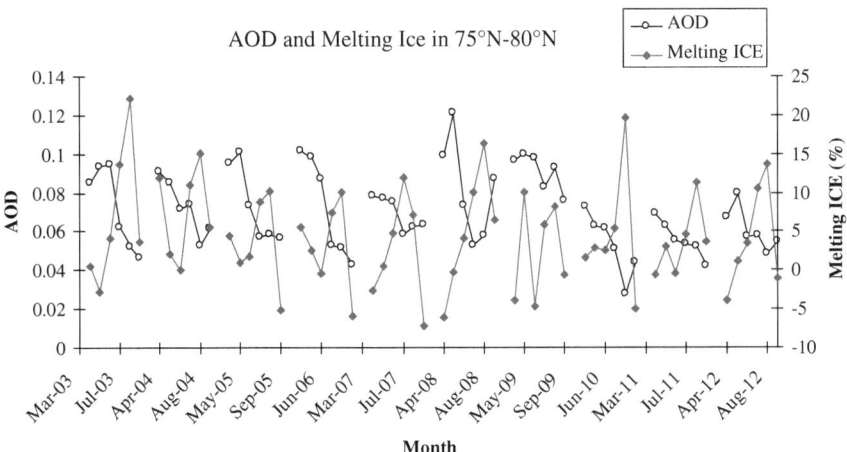

Fig. 3.8 Monthly mean MI and AOD time series in 75°N–80°N

Table 3.2 The peak time for AOD and ICE in 2003–2012 (unit:day)

	75–80°N	75–80°N
	AOD	ICE
2003	80	80
2004	96	72
2005	120	88
2006	120	80
2007	80	112
2008	80	112
2009	128	80
2010	104	88
2011	104	88
2012	128	144

Table 3.3 The lag time between AOD 与 ICE in 75°N–80°N (unit:month) (ICE ahead is positive)

2003	2004	2005	2006	2007	2008	2009	2010	2011	2012
0	0.8	1	1.33	−1.07	−1.07	1.6	0.53	0.53	−0.53

Next, we do the unit root check to see if they are stationary sequence.

As the unit root of critical values of AOD for level 1, 5, and 10 % are −3.540, −2.909, −2.592, respectively, they are all less than the t test value (−1.828); hence, AOD was non-stationary sequence.

We need to do first difference unit root checking for AOD. As the unit root of critical values for first difference unit root of AOD are −3.540, −2.909, and

Table 3.4 Lag analysis between ICE and AOD in 75°N–80°N

Variable	Coefficient	Std. error	t-Statistic	P value
C	0.0298	0.0209	1.4226	0.1599
ICE(-1)	0.0010	0.0003	3.4654	0.0010
ICE(-2)	0.0005	0.0003	1.5552	0.1251
ICE(-3)	-0.0005	0.0003	-1.4081	0.1642
ICE(-4)	-0.0004	0.0003	-1.4244	0.1594

Dependent variable: AOD, method: least square
$R^2 = 0.4517$
F-statistic: 12.5638
Prob (F-statistic): 0.0000

Table 3.5 The linear regression analysis between ICE(-1) and AOD in 75°N–80°N

Variable	Coefficient	Std. error	t-Statistic	P value
Constant (C)	-0.0115	0.01143	-1.0084	0.3169
ICE(-1)	0.0015	0.00023	6.5617	0.0000

Dependent variable: AOD, method: least square
$R^2 = 0.3912$
F-statistic: 43.056
Prob (F-statistic): 0.0000

-2.592 for level 1, 5, and 10 %, respectively, they are all greater than the t test value (-4.268). Hence, first difference unit root for AOD was stationary sequence.

The same rule can apply to the ICE. As the unit root of critical value of ICE for level 1, 5, and 10 % are all less than the t test, hence, ICE was non-stationary sequence, As the unit root of critical values for first difference unit root of ICE are -3.540198, -2.909206, and -2.592215 for level 1, 5, and 10 %, respectively, they are all greater than the t test value (-3.997949); hence, first difference unit root for ICE was stationary sequence.

We then need to see if AOD and ICE had co-integration relationship? Regression analysis for the both parameters is done to check whether their regression residuals are stationary. The least square regression models are used to do the regression for ICE(-1) respect to AOD. Then do the residual unit root-test. We found that the t test (-8.135) is smaller than all critical values (-2.599, -1.946, -1.614 for level 1, 5 and 10 %, respectively). The residual sequence had no unit root, and it was stationary sequence. Hence, AOD and ICE had co-integration relationship, and they had long-term equilibrium relationship.

3.2.3.3 The Correlation and Regression Analysis for AOD and CLD

Figure 3.9 is the AOD and cloud cover (CLD) time series for the 10 years (from March to September). It shows the negative relationship between them. The overall correlation coefficient was -0.233. AOD was 1 month lagged behind CLD. After shifting the peak time to the same time, there was a correlation relationship between them.

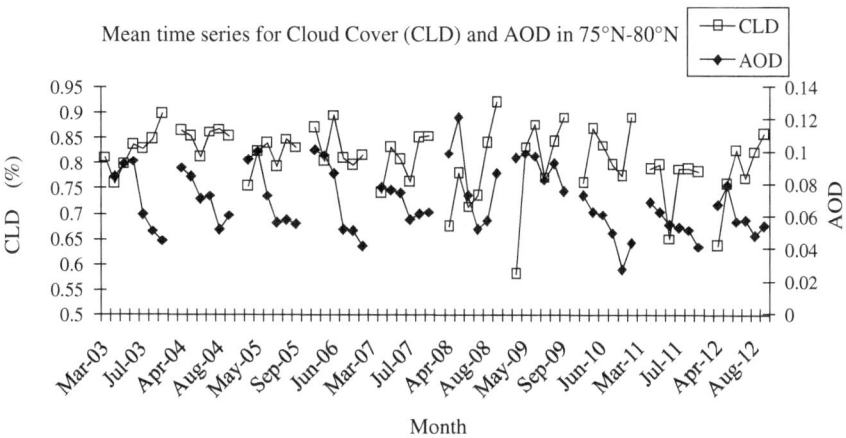

Fig. 3.9 AOD and CLD time series for year 2003–2012 in 70°N–80°N

It was reported that when calculating AOD using a dry rather than a wet aerosol, it would give a strong signal. In this case, the relationship becomes the AOD and CLD would have a strong negative correlation, the aerosol–cloud is formed by a wet precipitation process may remove a significant negative correlation between them (Quaas et al. 2010).

After shifting CLD 1 month behind, the correlation coefficient between CLD and AOD would change from 0.078 to −0.3929. Hence, AOD and CLD were less well correlated comparing to AOD and ICE. In year 2012, the coefficient would be −0.576. For precisely calculating the correlation coefficient between AOD and CLD, we do regression and lag regression analysis using EViews. We shift CLD 1, 2, 3, and 4 months behind, to see the correlations between CLD and AOD (Table 3.6).

CLD (−1) had the best results. The P value is the smallest (0.002 < 0.05). This result is coincident with the previous observation (CLD was 1 month ahead of AOD).

Table 3.7 shows CLD (−1) had influence to AOD, and as the P value of CLD(−1) is 0.0009 (<0.05), the correlation between AOD and CLD(−1) is significant. We do the unit root checking for CLD.

Table 3.6 The lag regression for CLD and AOD in 75°N–80°N

Variable	Coefficient	Std. error	t-Statistic	P value
C	0.1579	0.07978	1.9799	0.0522
CLD(−1)	−0.1797	0.05567	−3.2270	0.0020
CLD(−2)	−0.02198	0.05520	−0.3982	0.6919
CLD(−3)	0.02225	0.05514	0.4035	0.6880
CLD(−4)	0.05741	0.05572	1.0304	0.3069

Dependent variable: AOD, method: least square
$R^2 = 0.1790$
F-statistic: 3.3254
Prob (F-statistic): 0.0158

Table 3.7 Linear regression analysis result for CLD(-1) and AOD in 75°N–80°N

Variable	Coefficient	Std. error	t-Statistic	P value
Constant (C)	0.2057	0.04152	4.9535	0.0000
CLD(-1)	-0.1798	0.05148	-3.4918	0.0009

Dependent variable: AOD, method: least square
$R^2 = 0.1540$
F-statistic: 12.193
Prob (F-statistic): 0.00085

As t test statistical value is -6.5872, it is less than the 3 critical values of 3 levels in Table 3.8. Hence, CLD sequence did not have unit root. That means it was the stable sequence.

Now, we test if AOD and CLD had co-integration relationship. The regression analysis is further carried out and followed by checking the smoothness of the regression residuals.

Table 3.9 shows the residual sequence has no unit root, and it was the stationary sequence (-3.558511 is less than the three critical values). Hence, the first-order difference of AOD and CLD had co-integration, and they had a long-term equilibrium relationship.

We have found AOD lagged CLD 1 month behind. There were certain negative correlations between CLD and AOD. That means the increasing of CLD would reduce the AOD content in some degree. On the other hand, reducing CLD would increase AOD relatively. That explained there were less CLD and more AOD in April.

The correlation coefficients between AOD and CLD after shifting are -0.935, -0.548, and -0.548, respectively, for years 2003, 2009, and 2012. Year 2003 had very high negative correlations between CLD and AOD.

Generally, AOD and CLD had less degree of correlation comparing the correlation between AOD and ICE. That means the Arctic aerosols are mainly from seawater rather than from the high cloud.

Table 3.8 Unit root checking for CLD

Augmented Dicky-Full test statistic		t-Statistic	Prob.
		-6.5872	0.0000
Test critical values	1 % level	-3.5300	
	5 % level	-2.9048	
	10 % level	-2.5899	

Null hypothesis: CLD has a unit root

Table 3.9 Unit root-test for the residuals between AOD and CLD

Augmented Dicky-Full test statistic		t-Statistic	Prob.
		-3.5585	0.0095
Test critical values	1 % level	-3.5402	
	5 % level	-2.9092	
	10 % level	-2.5922	

Table 3.10 Regression analysis among AOD, ICE, and CLD

Variable	Coefficient	Std. error	t-Statistic	P value
Constant (C)	−0.0614	0.0347	−1.7681	0.0824
CLD	0.0978	0.0363	2.6909	0.0093
ICE	0.0012	0.0002	6.1173	0.0000

Dependent variable: AOD, method: least square
$R^2 = 0.3987$
F-statistic: 18.899
Prob (F-statistic): 0.00001

3.2.3.4 The Relationship Among AOD, ICE, and CLD

To find the relationships among AOD, ICE, and CLD, a linear model needs to be set up. We still focus on the region 75°N–80°N.

The relationship equation is as follows:

$$AOD = -0.061414 + 0.097762\, CLD + 0.001158\, ICE\ (75°N\text{-}80°N) \quad (3.6)$$

In Table 3.10, goodness of fit $R^2 = 0.3987$. The t test and F test values were all refusing the original hypothesis. P values of ICE and CLD were all less than 0.05. That means Eq. (3.4) is significant. Both ICE and CLD all had influence on AOD.

There was a positive relationship between AOD and ICE and less significant correlations between CLD and AOD. However, there were other factors which could be the external driving forces to AOD, hence to the radiative balances. The Arctic anthropogenic aerosol concentration and composition, the commercial shipping through Arctic when sea ice is reduced, and the black carbon aerosol on the snow and ice–albedo are the other three driving forces for AOD. Other reasons could due to the response of DMS cycle to changes in sea-ice cover and further more to the arctic marine biology and the large scale ocean circulation's changes.

3.3 Conclusions

The distributions and correlation analysis between AOD and ICE, AOD and MI (MI), and AOD and CLD (cloud cover) are all studied. We focus the region on 75°N–80°N for correlation analysis where ice melted more. We found out that NAO and MI had significant influence on CHL.

Different from CHL with peak in June in generally, AOD was higher in spring and lower in summer. Year 2009 had higher AOD, and year 2010 had lower AOD. CLD was generally overcastted and with April less overcastted and more fluctuations.

ICE was generally 1 month ahead of AOD, and there were good correlation between ICE(−1) and AOD. AOD and MI would have negative relationship during early spring and summer. AOD and CLD also had negative relationship, with AOD 1 month lagged behind CLD. Both ICE and CLD all had influence on AOD. However, ICE had more impact on AOD than CLD. The correlation was higher for ICE comparing to AOD. That means AOD had less effect coming from cloud cover and more effect from sea-ice aerosols. Besides, AOD and ICE, AOD and CLD all had long-term equilibrium relationship.

References

Feingold, G., Eberhard, W. L., Veron, D. E., & Previdi, M. (2003). First measurements of the Twomey indirect effect using ground-based remote sensors. *Geophysical Research Letters, 30*, 1287.

Gabric, A. J., Qu, B., Matrai, P. A., & Hirst, A. C. (2005). The simulated response of dimethyl-sulphide production in the arctic ocean to global warming. *Tellus, 57B*, 391–403.

Kaufman, Y. J., Koren, I., Remer, L. A., Rosenfeld, D., & Rudich, I. (2005). The effect of smoke, dust, and pollution aerosol on shallow cloud development over the Atlantic Ocean. *Proceedings of National Academy of Science, 102*, 11207–11212.

Monahan, E. C., & Muircheartaigh, I. (1980). Optimal power-law description of oceanic white-cap coverage dependence on wind speed. *Journal of Physical Oceanography, 10*, 2094–2099.

Nilsson, E. D., Martensson, E. M., Ekeren, J Sv, Leeuw, Gd, Moerman, M. M., & O'Dowd, C. (2007). Primary marine aerosol emissions: Size resolved Eddy covariance measurements with estimates of the sea salt and organic carbon fractions. *Atmospheric Chemistry and Physics Discussions, 7*, 13345–13400.

Pang, H. (2007). *Econometrics* (pp. 265–284). Beijing: Science Publishing Press.

Quinn, P. K., Shaw, G., Andrews, E., Dutton, E. G., Ruoho-Airola, T., & Gong, S. L. (2007). Arctic haze: Current trends and knowledge gaps. *Tellus B, 59*, 99–114.

Quinn, P. K., Bates, T. S., Baum, E., Doubleday, N., Fiore, A. M., Flanner, M., et al. (2008). Short-lived pollutants in the Arctic: Their climate impact and possible mitigation strategies. *Atmospheric Chemistry and Physics, 8*, 1723–1735. doi:10.5194/acp-8-1723-2008.

Ramanathan, V., Ramana, M. V., Roberts, G., Kim, D., Corrigan, C., Chung, C., et al. (2007). Warming trends in Asia amplified by brown cloud solar absorption. *Nature, 448*, 575–578.

Serreze, M. C., Holland, M. M., & Stroeve, J. (2007). Perspectives on the Arctic's shrinking sea-ice cover. *Science, 315*(5818), 1533–1536. doi:10.1126/science.1139426.

Serreze, M., Walsh, J., Chapin, F., Osterkamp, T., Dyurgerov, M., Romanovsky, V., et al. (2000). Observational evidence of recent changes in the northern high latitude environment. *Climate Change, 46*, 159–207.

Slingo, A. (1990). Sensitivity of the Earth's radiation budget to changes in low clouds. *Nature, 343*, 49–51.

Struthers, H., Ekman, A. M. L., Glantz, P., Iversen, T., Kirkevåg, A., Mårtensson, E. M., et al. (2011). The effect of sea-ice loss on sea salt aerosol concentrations and the radiative balance in the Arctic. *Atmospheric Chemistry and Physics, 11*, 3459–3477.

Chapter 4
Photosynthetically Active Radiation, Ice Cover, and Sea Surface Temperature

Abstract Relationships among photosynthetically active radiation (PAR), ice cover (ICE), sea surface temperature (SST), and mixed layer depth (MLD) were studied. Our research area is in Greenland Sea (20°W–10°E, 65°N–85°N), and time scale is during year 2003–2012. Generally, PAR exhibited as a normally distribution. PAR began to rise from March and reached to peak in June, then gradually declined. MLD was higher in spring and autumn and lower in summer. The regression analysis for PAR and ICE, PAR and SST were carried out. PAR had lag correlation with ICE, and ICE was lagged 2 months ahead of PAR. ICE had a significant effect on PAR. ICE and PAR also had a long-term equilibrium relationship. The peak of PAR was 1 month behind of melting ice (MI) in southern region (65°N–70°N) and 1 month ahead of MI in north of 70°N. SST lagged PAR 1 month behind, and PAR has a significant effect on SST. PAR had a decreasing trend from 2010 to 2012. There was strong correlation between PAR and SST.

Keywords Photosynthetically active radiation (PAR) · Ice cover (ICE) · Melting ice (MI) · Mixing layer depth (MLD) · Sea surface temperature (SST)

4.1 Introduction

The research on PAR and ICE

PAR is the solar energy in the spectrum of 400–700 nm that photosynthetic organisms are able to use in the process of *photosynthesis*. This spectral region corresponds with the range of light visible to the human eye McCree (1981). It is an important forcing of photosynthesis. CHL, the most abundant plant pigment, is most efficient in capturing red and blue light. *Accessory pigments* harvest some green light and pass it on to the photosynthetic process (http://en.wikipedia.org/wiki/Photosynthetically_active_radiation). The phytoplankton biomass is influenced by the incident PAR. It is a key variable in ecosystem models. The growth of all living things needs their nourishment from photosynthesis, so does the growth of phytoplankton biomass. During photosynthesis process,

© The Author(s) 2015   49
B. Qu, *The Impact of Melting Ice on the Ecosystems in Greenland Sea*,
SpringerBriefs in Environmental Science, DOI 10.1007/978-3-642-54498-9_4

the sunlight contributes to radiative energy, which is then converted to chemical energy by using atmospheric CO_2 (http://en.wikipediawiki/Carbon_dioxide_in_ Earth%27s_ atmosphere).

PAR plays an important role in photosynthesis and hence in net primary production. The temperature, light, nutrients, and pigment content in algae are the most important factors controlling PAR. The ice algae have great relationship with PAR. Finenko et al. (2002) tried to find the combination factors (physical, chemical, and biological) on the changing of P_{max}^B and α^B, where P_{max}^B is the maximum intensity of photosynthesis and α^B is the initial slope of the light curve; they found that the key factor for determining P_{max}^B is temperature, while the α^B is dependent on the nitrate and chlorophyll concentrations and also on the light absorption by phytoplankton. The reducing of CHL concentrations is due to the partial absorption of the light of limited photosynthetic utility.

Fritsen et al. (2010) studied the timing of sea-ice formation and exposure to PAR in the Antarctic Peninsula. They found that the timing of ice formation and ice melting had significant impact on the distribution of time integrated exposure to PAR (TIEP). The availability of PAR also has great influence on the biomass accumulation during ice formation and through the ice-melting process (Fritsen et al. 2010). Arrigo et al. (1997) suggested that the light limitation in autumn could be the primary determinant of algal growth. It was reported that, for a clear day, the growth ratio of PAR and solar irradiance was due to increase in AOD (Aculinin 2007).

Data and methods are listed in Sect. 2.2. Here, we focus on the relationship between PAR, ICE, and SST, and will also look insight into the relationship between PAR and MLD to determine how ice cover influences PAR and other parameters.

4.2 Results

4.2.1 The Distributions of PAR

Ten-year mean PAR distribution with standard deviations in the study region (65–85°N, 20°W–10°E) is shown in Fig. 4.1. It exhibited a normal distribution with peaks around June. The short standard deviation bar tells us that the distribution is evenly distributed and the differences between the years were small. Figure 4.2 shows the yearly mean PAR for the 10 years. Year 2008 and 2012 had higher PAR than other years. Eight-day mean for each year is shown in Fig. 4.3. Year 2003 had higher PAR in spring and June. Year 2008 had relative higher PAR in summer. Year 2009 had missing data in summer but exhibited the highest PAR in July. Year 2004 also had relative higher distributions in early March and late summer but lower in spring and early summer. Year 2010 had lower PAR in summer, and year 2012 had lower PAR in late summer and autumn.

We still divide the study region into 4 subregions with 5° zonal apart.

In 65°N–70°N (Fig. 4.4a), Year 2008 had lower PAR in April and higher PAR in May. PAR in 2011 had much lower value. PAR was consistently lower in year

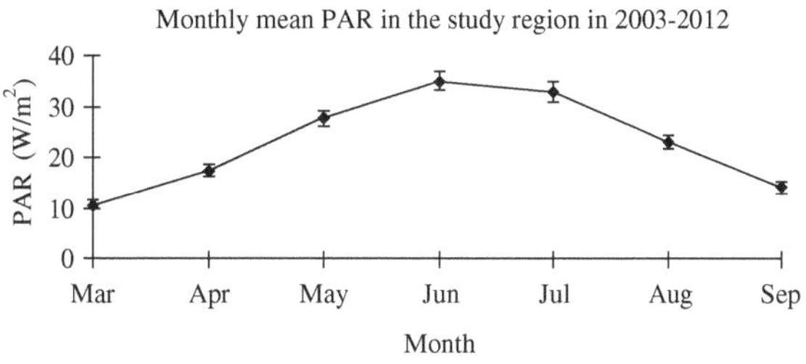

Fig. 4.1 10-year monthly mean PAR in the study region. Error bars are the standard deviations

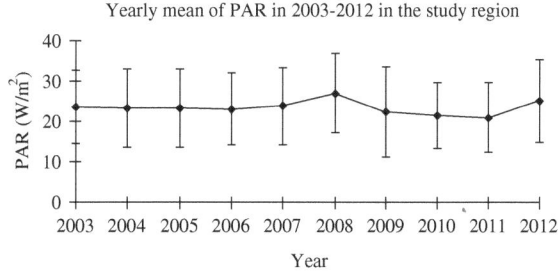

Fig. 4.2 Yearly mean PAR for the 10 years in the study region (65°N–85°N, 20°W–10°E)

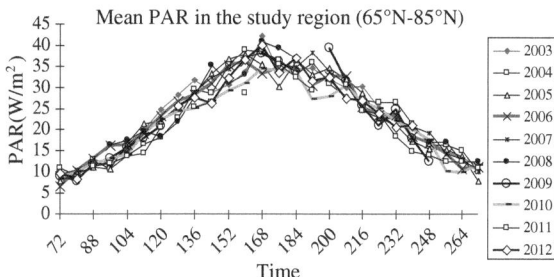

Fig. 4.3 Eight-day mean PAR in each year for the 10 years in the study region (65°N–85°N, 20°W–10°E)

2010 from spring throughout summer. Year 2007 had higher peak in May, and year 2012 had relative higher PAR in June. Year 2004 had higher PAR from middle of July to August. Year 2003 had a dip in middle May but rose to the high peak within one month. Year 2009 had higher PAR in summer although the data were missing before July.

In 70°N–75°N (Fig. 4.4b), year 2008 had low value in early May but raised sharp and reached to a higher peak in middle of May. Year 2010 had lower values from June and September with a dip in June. Year 2005 had higher values in April through early May but low in middle of May.

In 75°N–80°N (Fig. 4.4c), year 2011 had low PAR in early spring and reached to the highest in June, then to the lowest in autumn. There were some missing data in August for two weeks. Year 2009 had peak in June but data missing in earlier time. Year 2007 had higher value in June and early July. Year 2004 had low spring values and relative higher values in May. Year 2010 had relative lower PAR especially in May, but had higher PAR after August.

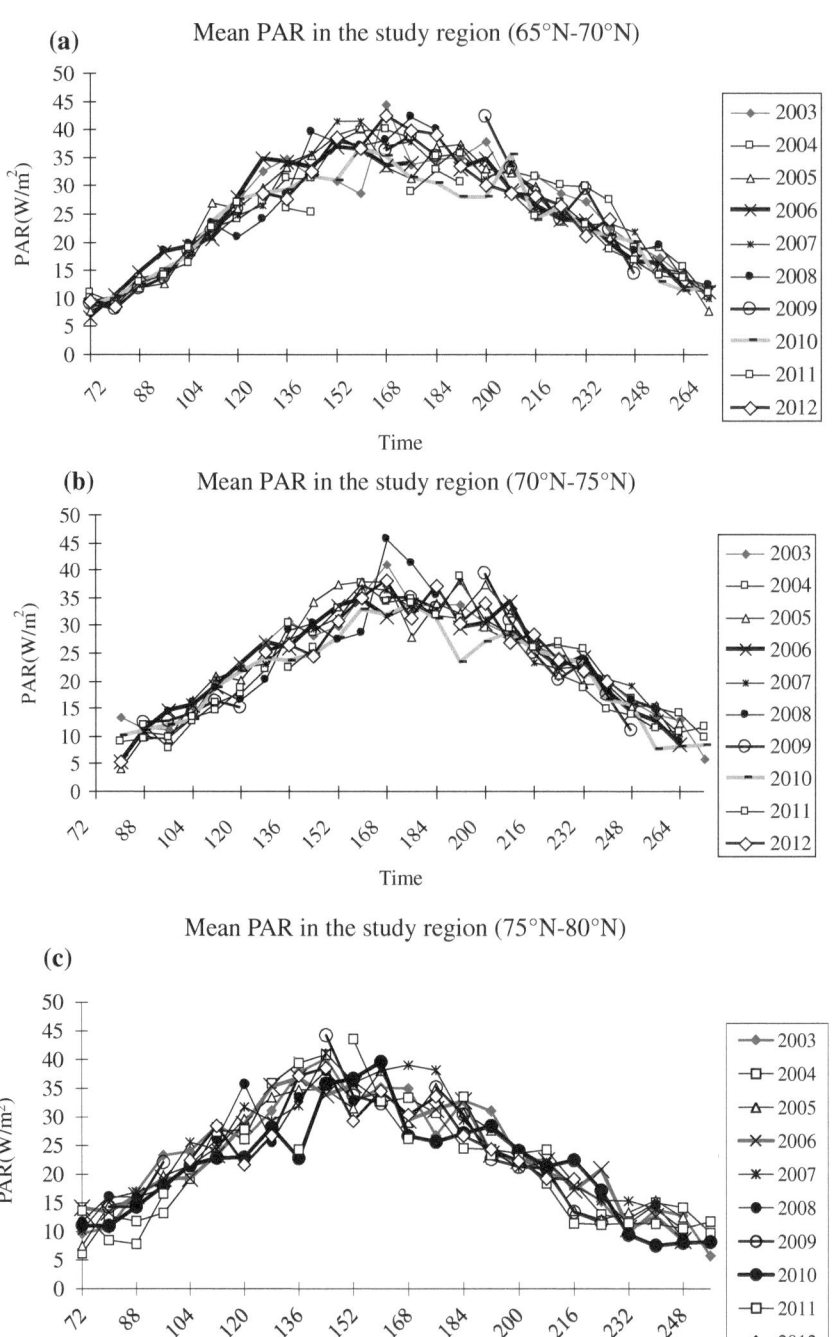

Fig. 4.4 Eight-day mean PAR in the subregions (65°N–80°N, 20°W–10°E)

Zonal PAR distribution is in Fig. 4.5 with much higher PAR in year 2012 and much lower in year 2011. Generally, PAR was higher in south and lower in north with a sharp decrease around north of 80°N. It is interesting to see there was a small peak around 80°N especially for year 2012. Year 2010 had relative constant lower PAR along latitude.

The PAR in three dimensions for years 2009 and 2010 is shown in Fig. 4.6. It shows north lower and south higher distributions. For the same latitude up north, the distribution was east higher and west lower. But in south of 80°N, there was no much difference from east to west. Year 2010 had smoother and lower distribution compared to year 2009.

4.2.2 Mixed Later Depth (MLD) Distributions

Monthly mean MLD is also calculated and shown in Fig. 4.7. Due to the special geographical location, the various (standard deviations) are large. Generally, it was higher in spring and autumn and lower in summer.

Mean MLD averaged along latitude 20°W–10°E is in Fig. 4.8. Generally, it was higher in winter and spring and lower in summer and increased gradually from autumn. The unusual peaks in May in 74.5°N and 75.5°N are due to the special geographic location in the region. There is an anticlockwise circulation current pattern within 71°N–75°N. There is a pit located around 74°N–75°N where middle of the circulation locates. The MLD is much smaller in north of 77°N.

Mean MLD along 4 subregions with 5° zonal apart is shown in Fig. 4.9. 70°N–75°N is a region with higher MLD located especially in spring. It is interesting that within northern band (80°N–85°N), MLD is lower in spring and higher in summer.

Monthly 3D distributions from March to July are shown in Fig. 4.10. From west to east, MLD is higher in the both sides and lower in the middle. MLD reaches to its minimum by June and after that, it gradually increases.

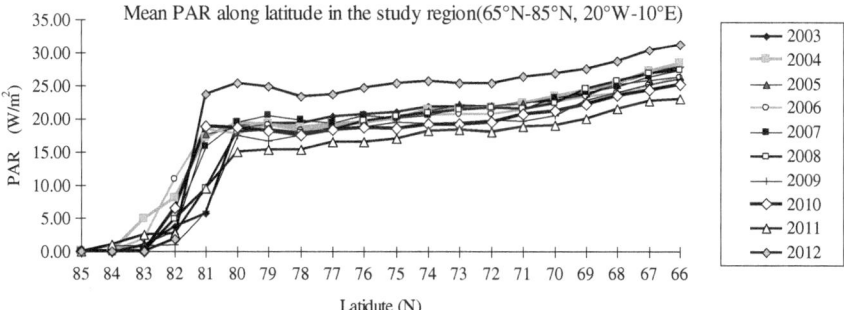

Fig. 4.5 Zonal mean PAR distribution for in the study region (20°W–10°E)

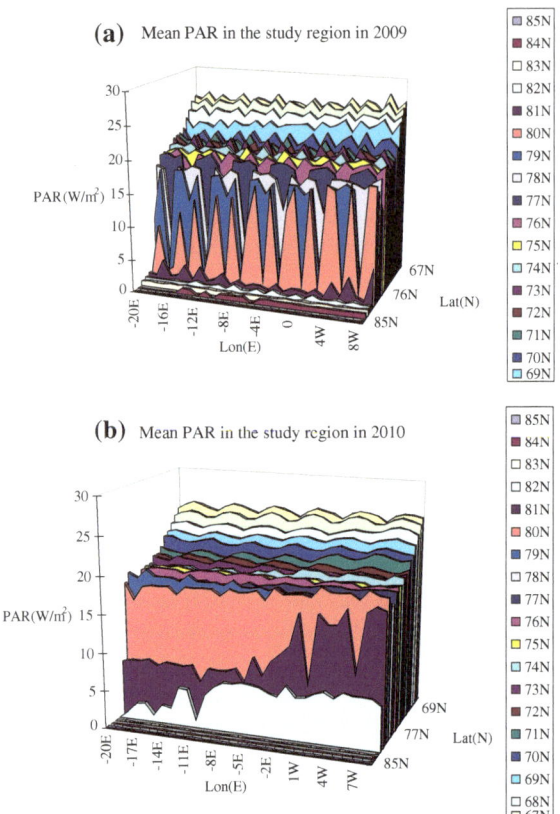

Fig. 4.6 3D yearly mean PAR in year 2009 in the study region

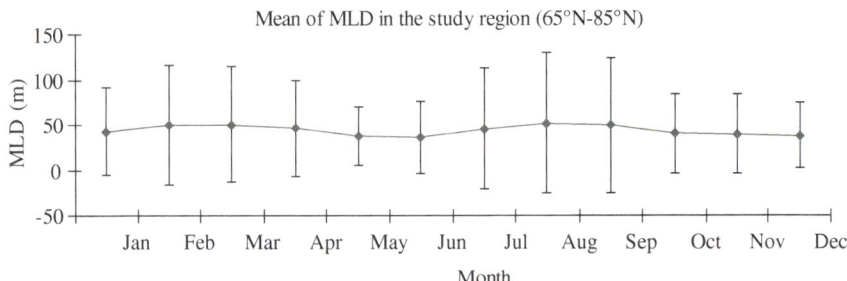

Fig. 4.7 Monthly mean MLD in the study region. Vertical bars are the standard deviations

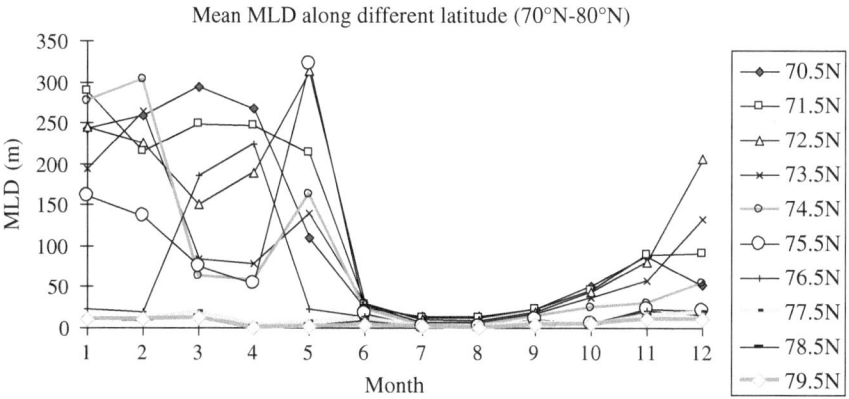

Fig. 4.8 Monthly mean MLD along longitude (20°W–10°E)

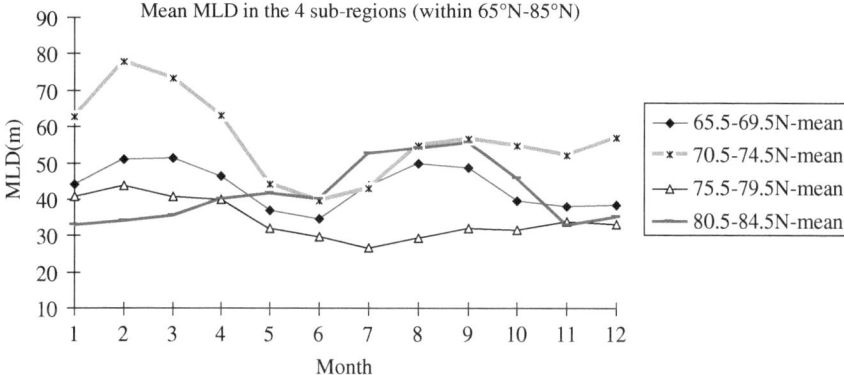

Fig. 4.9 Monthly mean MLD in different 5° zonal bands within 20°W–10°E

4.2.3 The Correlation Analysis for PAR and ICE

The correlation coefficients between PAR and ICE are shown in Fig. 4.11. Generally, there was a negative correlation between PAR and ICE. That means the reduction of ICE would increase PAR content. However, there were positive correlations in 75°N–80°N apart from year 2007. In the recent years (after 2010), the correlations tended to be positive north of 70°N. There was a strong negative correlation in year 2009 in southern region.

Figure 4.12a–d shows 10-years time series for PAR and MI. In general, the peak of PAR was 1 month behind MI in southern region (65°N–70°N) and PAR was ahead of MI in north of 70°N, although things would change from time to time. Sometimes, both peaks would meet at the same time.

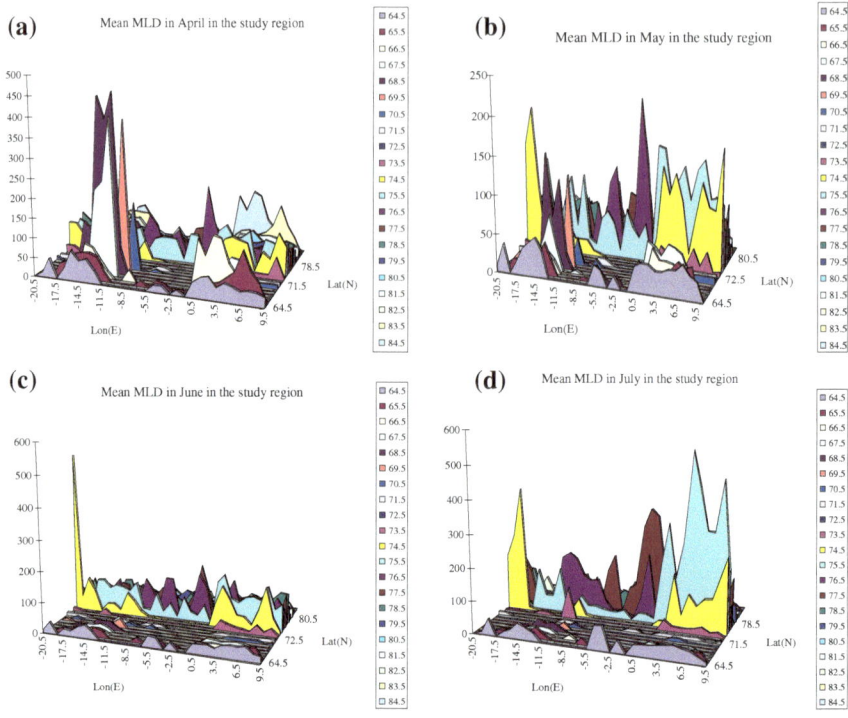

Fig. 4.10 Monthly MLD 3D distributions from April to July

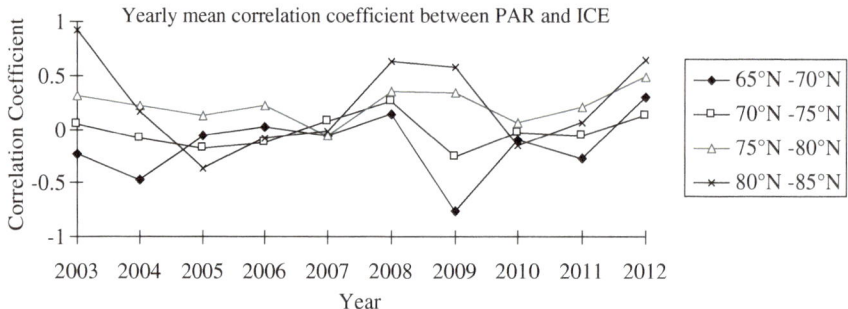

Fig. 4.11 Yearly mean correlation coefficients between PAR and ICE

Figure 4.13 shows that PAR and ICE had a relationship with a time lag in the whole study region (65°N–85°N). PAR started to increase with the increase in ICE. After a while, ICE decreased with increase in PAR. After a few months, ICE

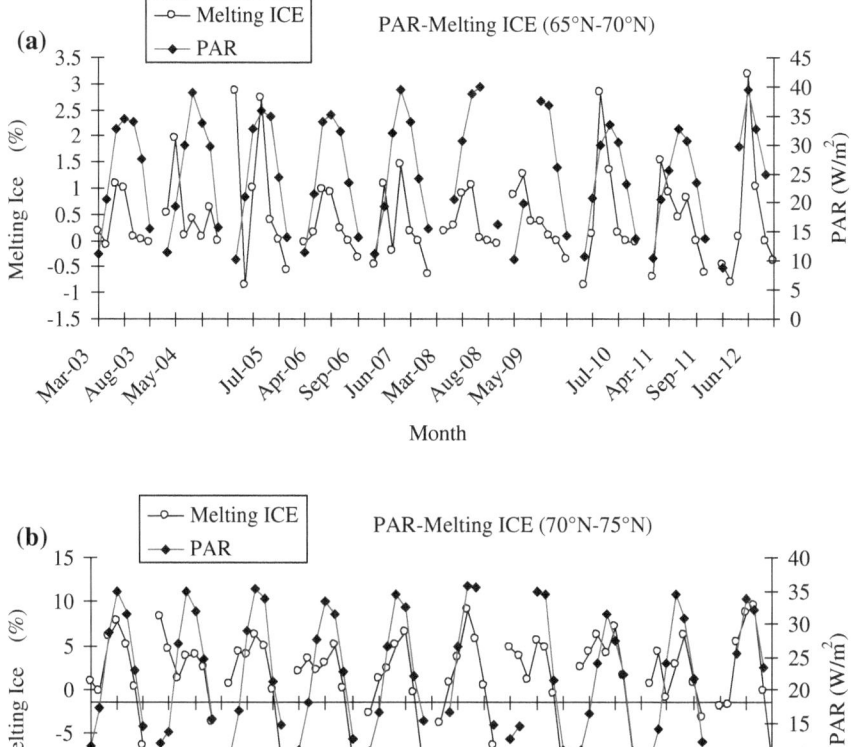

Fig. 4.12 Time series of PAR and MI in the 4 subregions

further decreases with the increase in PAR. Hence, the peak time of PAR and ICE was different. The time lag table is shown in Table 4.1.

ICE ahead of PAR shows positive value. The time lag between PAR and ICE is listed in Table 4.2.

Table 4.3 indicated the time lag of PAR and ICE. Generally, year 2012 had the shortest time lag and year 2009 had the longest time lag in southern regions. Years 2010 and 2005 had the longest time lag in the northern regions. Year 2006 had unusual lags with PAR ahead of ICE in northern regions. In general, ICE was about 2 months ahead of PAR in the study region.

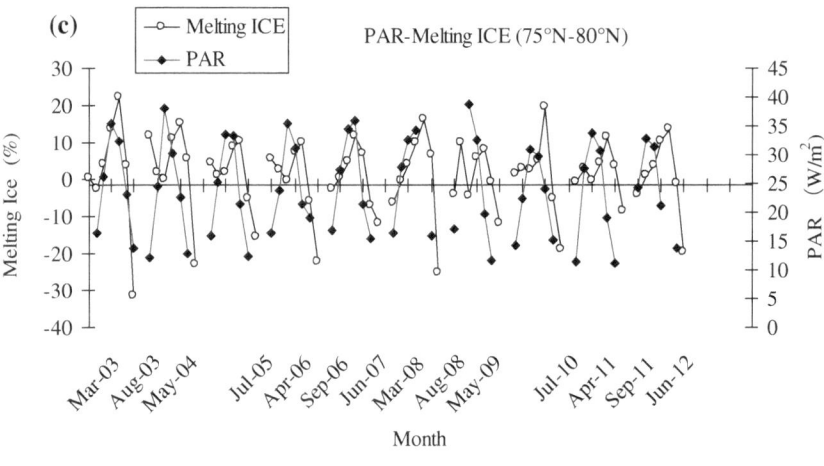

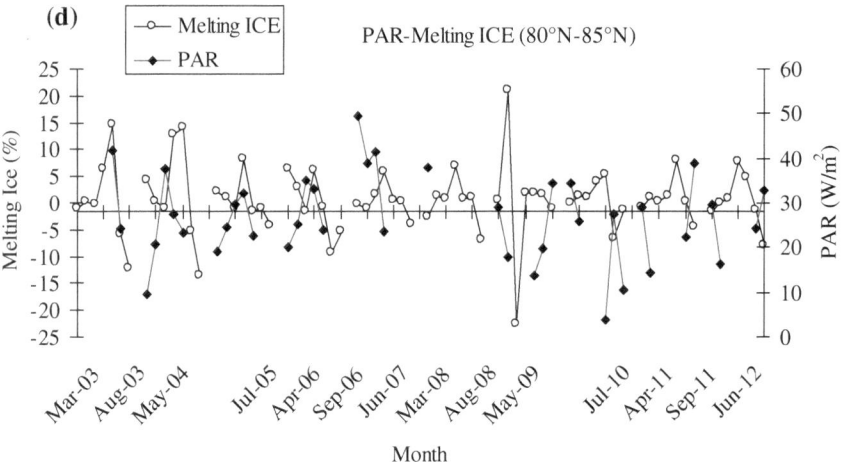

Fig. 4.12 (continued)

We use EViews (Pang 2007) to confirm the time lags. We denote ICE(-1), ICE(-2), ICE(-3), and ICE(-4) as shifting ICE 1, 2, 3, 4 months behind, respectively.

4.2.4 Regression and Lag Analysis for PAR and ICE

EViews statistical software was used for regression analysis (Pang 2007). The regressions for PAR and ICE after shifting in each subregion are shown in Tables 4.3, 4.4 and 4.5. Here, we ignore the most northern part of the subregion (80°N–85°N) due to less data available.

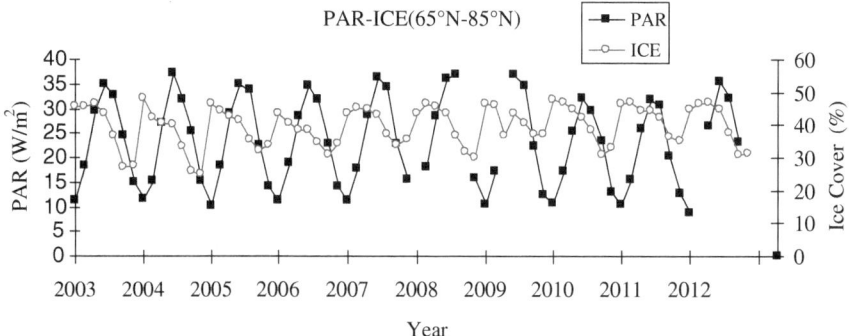

Fig. 4.13 Time series of PAR and ICE in the study region

Table 4.1 Peak time table (Unit: Day)

	65°N–70°N		70°N–75°N		75°N–80°N			80°N–85°N
	PAR	ICE	PAR	ICE		PAR	ICE	PAR
2003	168	80	168	80	2003	168	80	168
2004	168	104	192	72	2004	168	104	192
2005	160	72	168	88	2005	160	72	168
2006	152	120	160	80	2006	152	120	160
2007	152	96	192	112	2007	152	96	192
2008	176	112	168	112	2008	176	112	168
2009	200	80	200	80	2009	200	80	200
2010	160	112	176	88	2010	160	112	176
2011	160	88	160	88	2011	160	88	160
2012	168	160	168	114	2012	168	160	168

The correlation coefficients have been improved a lot especially for ICE(-2) in all subregions. The p value (Prob.) are all the smallest (0.00000 < 0.005) for ICE (-2), which means it was significantly correlated between PAR and ICE(-2). The results agree with the previous analysis (Table 4.2).

We focus on the region 75°N–80°N to see whether the two parameters PAR and ICE had co-integration (Table 4.6).

Table 4.7 shows that PAR and ICE(-2) had significant correlations. Next, we do the unit root-test for the PAR and ICE(-2) (Table 4.8).

Under 1, 5 and 10 % 3 significant level, Mackinnon critical values for unit root-test are −3.5348, −2.9069, and −2.5910, and t test value was −7.5365 which is less than the critical value, which means ICE(-2) series did not have the unit root; it was the stationary sequence. The same is reason for unit root-test of PAR, Table 4.9 shows the PAR also did not have the unit root, and it was a stationary sequence.

Next, to confirm whether there is a co-integration between ICE(-2) and PAR, we check the smooth regression residuals. PAR was set as being explanatory variable and ICE(-2) as the explanatory variable. We use ordinary least squares (OLS)

Table 4.2   The time lag between PAR and ICE (Unit: Day)

	65°N–70°N	70°N–75°N	75°N–80°N	80°N–85°N
2003	88	88	40	40
2004	64	120	96	96
2005	88	80	104	104
2006	32	80	−112	−104
2007	56	80	56	8
2008	64	56	56	88
2009	120	120	56	40
2010	48	88	104	104
2011	72	72	88	120
2012	8	54	56	8
Mean	64	83.8	54.4	50.4
	2.13 (Month)	2.79 (Month)	1.81 (Month)	1.68 (Month)
Mean	2.105 (Month)			

ICE was ahead of PAR

Table 4.3   Correlation analysis for PAR and ICE in 65°N–70°N

PAR	Corr.	R^2	F	Prob.
ICE	−0.175	0.031	1.99	0.163
ICE(-1)	0.468	0.219	17.35	0.000098
ICE(-2)	0.719	0.517	65.23	0.000000
ICE(-3)	0.486	0.236	18.54	0.000062
ICE(-4)	−0.101	0.010	0.612	0.437

Table 4.4   Correlation analysis for PAR and ICE in 70°N–75°N

PAR	Corr.	R^2	F	Prob.
ICE	−0.130	0.017	1.083	0.302
ICE(-1)	0.530	0.281	24.223	0.000007
ICE(-2)	0.810	0.657	116.745	0.000000
ICE(-3)	0.532	0.283	23.662	0.000009
ICE(-4)	−0.180	0.032	1.971	0.166

Table 4.5   Correlation analysis for PAR and ICE in 75°N–80°N

PAR	Corr.	R^2	F	Prob.
ICE	0.268	0.072	4.169	0.046
ICE(-1)	0.409	0.167	10.828	0.0017
ICE(-2)	0.630	0.397	34.934	0.0000
ICE(-3)	0.413	0.170	10.686	0.0002
ICE(-4)	−0.299	0.090	5.0208	0.029

Table 4.6 Peak time and time lag for PAR and ICE in 75°N–80°N (Unit:Day)

	PAR	ICE	Time lag
2003	168	128	40
2004	168	72	96
2005	184	80	104
2006	160	272	−112
2007	168	112	56
2008	168	112	56
2009	168	112	56
2010	184	80	104
2011	176	88	88
2012	168	112	56
Mean			54.4
			1.8133 (Month)

Table 4.7 The regression results for PAR and ICE(-2)

Variable	Coefficient	Std. error	t-Statistic	p value
C	0.4151	4.1401	0.1003	0.9205
ICE(2)	0.4757	0.0805	5.9105	0.0000

Dependent variable PAR, Method least square
$R^2 = 0.3973$
F-statistic: 34.9337
Prob. (F-statistic): 0.0000

Table 4.8 Unit root-test for ICE(-2)

		t-Statistic	Prob.
Augmented Dicky-Full test statistic		−7.5365	0.0000
Test critical values	1 % level	−3.5348	
	5 % level	−2.9069	
	10 % level	−2.5910	

Null Hypothesis D (ICE(-2)) has a unit root

Table 4.9 Unit root-test for PAR

		t-Statistic	Prob.
Augmented Dicky-Full test statistic		−8.5679	0.0000
Test critical values	1 % level	−3.6394	
	5 % level	−2.9511	
	10 % level	−2.6143	

Null Hypothesis PAR has a unit root

regression method (Pang 2007) to evaluate the regression model; furthermore, unit root-test for the residual was done (Table 4.10).

Under 5 % significant level's setting, t test statistical value is −8.5739, which is less than the 3 critical values. Hence, we refuse the original hypothesis. The residual sequence did not have unit root. It was the stationary sequence. Hence, there

Table 4.10 Unit root-test for the residual

		t-Statistic	Prob.
Augmented Dicky-Full test statistic		−8.5739	0.0000
Test critical values	1 % level	−2.6649	
	5 % level	−1.9557	
	10 % level	−1.6088	

Time Series for PAR and SST in 75°N-80°N

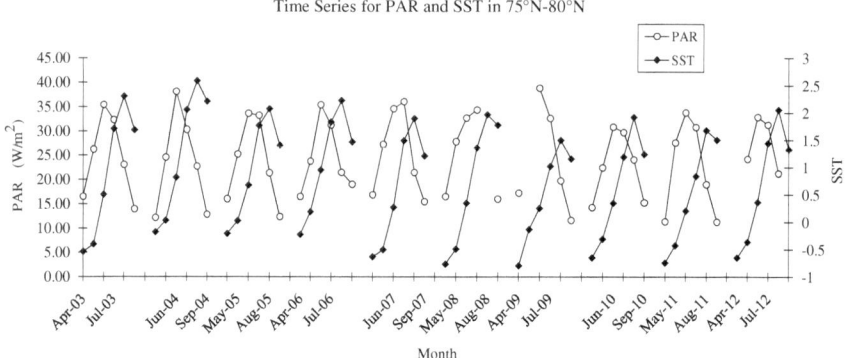

Fig. 4.14 Time series for PAR and SST in the region 75°N–80°N

Table 4.11 The lag analysis for PAR and SST in 75°N–80°N region

Variable	Coefficient	Std. error	t-Statistic	p value
C	22.2140	0.8032	27.6555	0.0000
SST	1.4953	0.6410	2.3328	0.0241
SST (1)	3.1911	0.6842	4.6641	0.0000
SST (2)	0.5228	0.6869	0.7612	0.4504
SST (3)	−2.8141	0.6627	−4.2462	0.0001
SST (4)	−2.3939	0.06816	−3.5119	0.0010

Dependent variable PAR, Method least square
$R^2 = 0.9145$
F-statistic: 98.4289
Prob. (F-statistic): 0.0000

was a co-integration between PAR and ICE(-2). That means they had long-term equilibrium relationship.

4.2.5 The Regression and Lag Analysis for PAR and SST

Figure 4.14 is the time series of SST and PAR in 75°N–80°N. There was a strong correlation between them. Usually, PAR was one month ahead of SST.

For 75°N–80°N region, the lag analysis for PAR and SST is listed in Table 4.10. It confirmed that the time lag between PAR and SST was 1 month and

Table 4.12 Regression equations and F test for PAR and SST

	Regression equation	F	R^2	Corr. coef. after shifting
70–75°N	PAR = 8.6951 + 4.1338SST	198.52	0.76	0.87
80–85°N	PAR = 19.014 + 6.4427SST	90.58	0.63	0.79

the influence of PAR to SST is significant. The goodness of fit ($R^2 = 0.9145$) is excellent (Table 4.11).

The regression equations for the other two regions (70°N–75°N and 80°N–85°N) are listed in Table 4.12. The correlations are still high (0.76 for 70°N–75°N and 0.63 for 80°N–85°N).

4.3   Conclusions

The distribution and correlation analysis between PAR and ICE, PAR and MI, and PAR and SST are all studied. We focus on the region 75°N–80°N for correlation analysis where ice melted more. PAR exhibits normal distributions with its peaks in June. Year 2008 had higher peak in early June, and 2010 had lower dip in later June. Year 2012 had highest PAR in the study region, and year 2011 had lowest PAR. In general, PAR in year 2010 kept low throughout the year.

Generally, MLD is higher in spring and autumn and lower in summer. The peak of PAR was 1 month behind MI in southern region (65°N–70°N) and 1 month ahead of MI in north of 70°N. Different from MI, ICE was about 2 months ahead of PAR. The influence of ICE to PAR was not that strong, and the goodness of fit was 0.4. PAR had strong correlation with SST with goodness of fit as high as 0.91 in 75°N–80°N and 0.87 in 70°N–75°N.

References

Aculinin, A. (2007). Diurnal Variation Of Aerosol Optical Properties Retrieved From Solar Radiation Measurements At Chisinau Site, Moldova. *Moldavian Journal of the Physical Sciences, 6*(N3–4), 392–397.

Arrigo, K. R., Lizotte, M. P., Worthen, D. L., Dixon, P., & Dieckmann, G. (1997). Primary production in Antarctic sea-ice. *Science, 276*, 394–397.

Finenko, Z. Z., Ya, T., Churilova, H., Sosik, M., & Basturk, O. (2002). Variability of photosynthetic parameters of the surface phytoplankton in the Black Sea. *Oceanology, 42*(1), 53–67.

Fritsen, C. H., Memmott, J. C., Ross, R. M., Quetin, L. B., Vernet, M., & Wirthlin, E. D. (2010). The timing of sea-ice formation and exposure to PAR during austral autumn and winter along the Western Antarctic Peninsula. *Polar Biol,* doi:10.1007/s00300-010-0924-7.

Pang, H. (2007). *Econometrics* (pp. 265–284). Beijing: Science Publishing Press.

Chapter 5
The Correlation Analysis and Predictions for Chlorophyll a, Aerosol Optical Depth, and Photosynthetically Active Radiation

Abstract The relationships between chlorophyll a (CHL), aerosol optical depth (AOD), and photosynthetically active radiation (PAR) were studied and predictions for the future 3 years are carried out. Our research area and timescale are in the Greenland Sea (20°W–10°E, 65°N–85°N) during the period 2003–2012. The regression analysis for CHL and AOD, CHL and PAR, and AOD and PAR were carried out. CHL and AOD had positive correlations (correlation coefficients were within 0.4–0.5) with AOD 2 months ahead of CHL. CHL and PAR were positively correlated and correlation coefficient was as high as 0.74 in 70°N–80°N. AOD was 2 months ahead of PAR and the goodness of fit were 0.63 and 0.42 for 70°N–75°N and 75°N–80°N, respectively. CHL, AOD, and PAR were all non-smooth sequences, with its first-order differential sequences were smooth and non-white noise time series. Therefore, the future 3 years prediction has been done using ARMA model.

Keywords Chlorophyll a (CHL) · Aerosol optical depth (AOD) · Photosynthetically active radiation (PAR) · ARMA model · Prediction · Regression

5.1 Introduction

Pelagic phytoplanktons are the major producers of organic matter in open waters, but sea-ice algae contribute additional production, particularly in late spring and early summer (Horner and Schrader 1982). Phytoplankton biomass in the Arctic is marked by high seasonality associated with light conditions, sea-ice cover, and nutrient availability (Harrison and Cota 1990). From a bottom-up view, the timing of phytoplankton production is set by light conditions and the magnitude of primary production by nutrient availability, while from a top-down view, match and mismatch of primary producers and grazers will largely determine the occurrence of pelagic retention versus export food webs (Carmack and Macdonald 2002). Light is the major limiting factor for phytoplankton production in Arctic waters with multi-year ice cover, so that the phytoplankton growth season is restricted to the ice-melting period in summer (Spies et al. 1988).

© The Author(s) 2015

B. Qu, *The Impact of Melting Ice on the Ecosystems in Greenland Sea*,
SpringerBriefs in Environmental Science, DOI 10.1007/978-3-642-54498-9_5

Aerosol optical depth (AOD) is a quantitative measure of the extinction of solar radiation by aerosol scattering and absorption between the point of observation and the top of the atmosphere. It is a measure of the integrated columnar aerosol load and the single most important parameter for evaluating direct radiative forcing. Gabric et al. (2002) found there is a strong relationship between CHL and AOD in the sub-Antarctic Southern Ocean. Later, Gabric et al. (2005) found that the coherence between remote sensed chlorophyll a (CHL) and AOD time series is strong in the band 50°–60°S, and there is a lag between the seasonal peaks which was suggested as being due to the emission of biogenic aerosol precursors from melting sea ice. The relationships among CHL, AOD, and ICE in smaller area (10°W–10°E, 70–80°N) in Greenland Sea are studied by Qu et al. (2014). Year 2009 was an unusual year in this particular region; CHL and AOD were much higher than other years especially in the northern part (75°N–80°N). The reason was explored and the driving forces were speculated to be the much increased ice melting and increased wind speed in autumn, plus increased deposit aerosol through the year.

Different from the results in Qu et al. (2014) where year 2009 was usual, here, in the larger area in Greenland Sea (extended from 10°W–10°E to 20°W–10°E), year 2010 was unusual and higher and early peaks appeared in this particular year. We have looked into the reasons caused for the peaks in 2010 especially in northern region 75°N–80°N (Chap. 3). This chapter looks into the relationships among CHL, AOD, and PAR in Greenland Sea. Our aim here is doing further correlation and regression study between CHL and AOD, AOD and PAR, PAR and CHL, and CHL and cloud cover (CLD) based on results from previous chapters. The prediction methods using ARMA model for CHL, AOD, and PAR are in use for the first time. The prediction data are obtained. Those data are used as an indication of the future trends. More accurate predictions are expected to be carried out in the future.

5.2 The Correlation Analysis for CHL, AOD, and PAR

5.2.1 The Correlation Analysis Between CHL and AOD

The data sources are shown in Chap. 3 (Sect. 3.2). We focused on the region 75°N–80°N. The peak times for CHL and AOD are calculated in Table 5.1.

Table 5.1 shows the time lag between CHL and AOD in 75°N–80°N subregion. The average time lag was around 2 months with AOD ahead of CHL. The time series for the CHL and AOD is shown in Fig. 5.1. AOD is ahead of CHL. There was a correlation between CHL and AOD.

After shifting CHL 2 months ahead, the correlation coefficients for the two regions are improved to 0.42 for the 70°N–75°N and 0.48 for the 75°N–80°N.

Table 5.2 shows the AOD(−2) had the smaller P value. That means AOD had significant influence on CHL by ahead of it 2 months' time. The result is consistent with the previous result. After shifting AOD 2 months behind, the regression analysis is carried out in Table 5.3.

Table 5.1 Peak time lag for CHL and AOD for year 2003–2012 in 75°N–80°N

	CHL (day)	AOD (day)	Time lag (month)
2003	144	80	2.133
2004	160	96	2.133
2005	184	120	2.133
2006	200	120	2.667
2007	168	80	2.933
2008	192	80	3.733
2009	160	128	1.067
2010	184	104	2.667
2011	160	104	1.867
2012	192	128	2.133
Mean	174.4	104	2.347

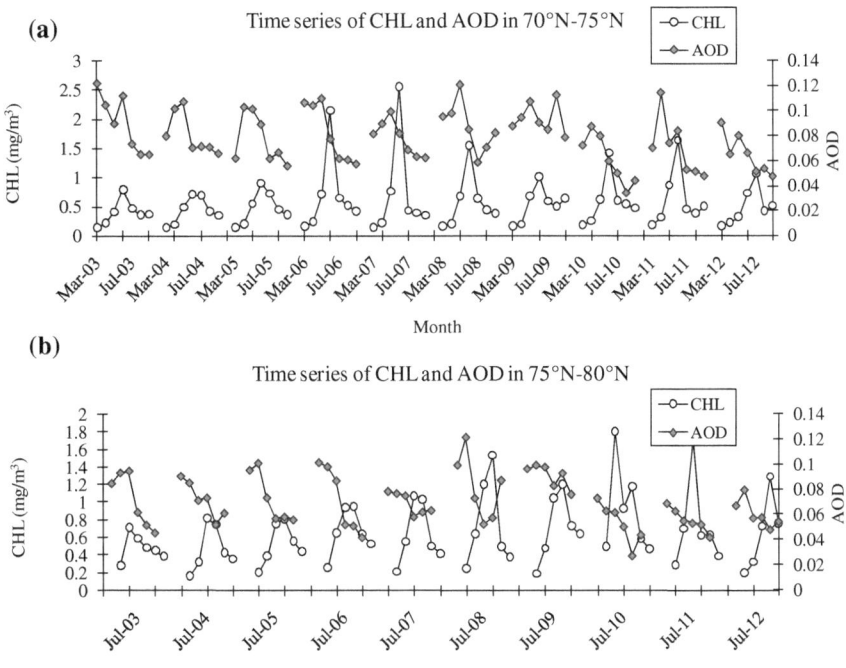

Fig. 5.1 Time series for the CHL and AOD for the 10 years 2003–2012 in the subregions **a** 70°N–75°N and **b** 75°N–80°N

The regression equation is:

$$CHL = -0.144 + 10.87 AOD \ (75°N-80°N) \tag{5.1}$$

Table 5.2 The lag analysis for CHL and AOD

Variable	Coefficient	Std. error	t-Statistic	P value
C	0.1174	0.2747	0.4272	0.6708
AOD(-1)	0.0106	2.7739	0.0038	0.9969
AOD(-2)	9.1967	3.1067	2.9603	0.0044
AOD(-3)	5.2377	3.0966	1.6915	0.0958
AOD(-4)	-8.1135	2.6786	-3.0290	0.0036

Dependent variable: CHL, Method: least square
$R^2 = 0.3674$
F-statistic: 8.8578
Prob (F-statistic): 0.00001

Table 5.3 The regression analysis for CHL and AOD after shifting (75°N–80°N)

Variable	Coefficient	Std. error	t-Statistic	P value
Constant (C)	-0.1437	0.1735	-0.8287	0.4114
AOD	10.8700	2.2825	4.7624	0.0000

Dependent Variable: CHL, Method: least square
$R^2 = 0.3255$
F-statistic: 22.6804
Prob (F-statistic): 0.000019

In Table 5.3, the goodness of fitting R^2 was 0.3255, under the significant level $\alpha = 0.05$. The t test and F test are carried out. The P value of AOD is smaller than 0.05. That means the regression equation is significant between CHL and AOD. The equation for 70°N–75°N is as follow:

$$CHL = -0.205 + 9.557AOD \, (70°N–75°N) \tag{5.2}$$

Next, the regression and residual test are done to see whether they are stationary sequences.

Table 5.4 shows the t-value is -5.7413 which was smaller than the three critical values under 1, 5, and 10 % level. That shows the residual sequence had no unit root, and it was stationary sequence. Hence, CHL and AOD had co-integration relationship. That means they had long-term equilibrium relationship.

5.2.2 Correlation Between CHL and Cloud Cover (CLD)

What is the relationship between CHL and CLD? Figure 5.2 shows the CHL and CLD time series for the southern region (70°N–75°N) and northern region

Table 5.4 Stationary test for estimating residuals for CHL and AOD

		t-Statistic	Prob.
Augmented dicky-full test statistic		-5.7413	0.0000
Test critical values	1 % level	-2.6120	
	5 % level	-1.9475	
	10 % level	-1.6126	

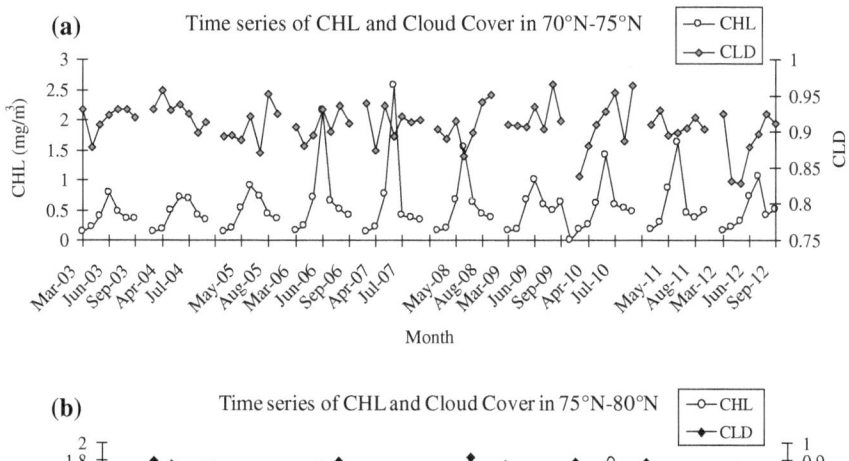

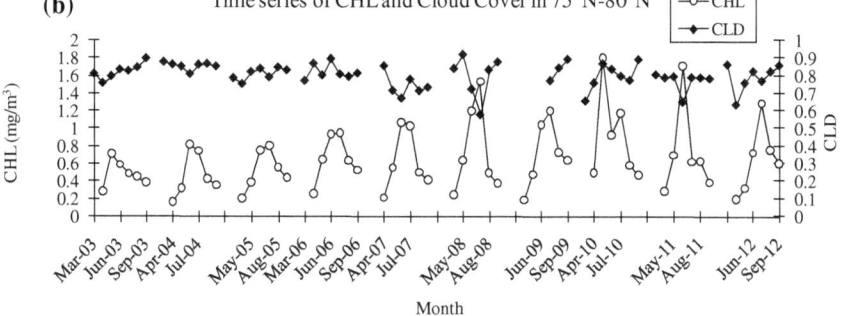

Fig. 5.2 Time series for the CHL and CLD for the 10 years 2003–2012 in the subregion **a** 70°N–75°N and **b** 75°N–80°N

(75°N–80°N). The correlation between CHL and CLD varied from years. In southern region, the higher correlation coefficients were 0.52 and 0.45 in year 2006 and 2010, respectively. In year 2008, the correlation coefficient was −0.55. CLD was 1 month behind of CHL in some years, but the tendency was not significant. In northern region, CHL and CLD had more negative correlations. The 10 years series had correlation coefficient of −0.2. CLD was 1 month ahead of CHL for most of years, apart from year 2006 (same peak) and 2012 (with CHL ahead of CLD 1 month). Year 2006 had good correlation of 0.62, and year 2007 and 2011 had negative correlation of −0.65 and −0.9, respectively. Due to Arctic Ocean had overcast cloud cover throughout the year, hence, CLD would have negative influence on CHL. More CLD led to less CHL productions.

5.2.3 Correlations Between CHL and PAR

Figure 5.3 is the time series of CHL and PAR for the region 70°N–75°N and 75°N–80°N.

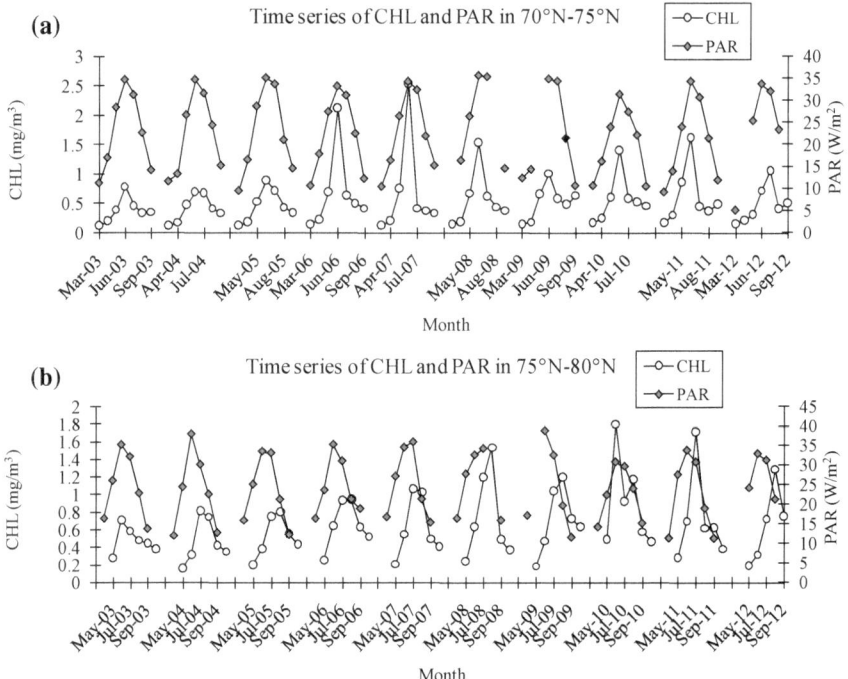

Fig. 5.3 Time series of CHL and PAR for the 10 years 2003–2012 in the subregion **a** 70°N–75°N and **b** 75°N–80°N

From Fig. 5.3, generally, CHL and PAR were positively correlated. The correlation coefficients were quite high. Table 5.5 listed the correlation coefficients for the three subregions. The correlation coefficients are as high as 0.65–0.75 in 75°N–80°N.

The regression analysis is shown in Table 5.6.

Table 5.5 Correlation coefficients between CHL and PAR for the three subregions

	70°N–80°N	70°N–75°N	75°N–80°N
Correlation coefficient	0.744	0.662	0.656

Table 5.6 Regression analysis for CHL and PAR (75°N–80°N)

Variable	Coefficient	Std. error	t-Statistic	P value
Constant (C)	−0.0664	0.1210	−0.5487	0.5855
ICE(−1)	0.0302	0.0047	6.3749	0.0000

Dependent Variable: CHL, Method: least square
$R^2 = 0.4294$
F-statistic: 40.6400
Prob (F-statistic): 0.0000

Table 5.7 Stationary test for estimating residuals for CHL and PAR (75°N–80°N)

		t-Statistic	Prob.
Augmented dicky-full test statistic		−5.0907	0.0000
Test critical values	1 % level	−2.6186	
	5 % level	−1.9485	
	10 % level	−1.6121	

The regression equation is as follow:

$$CHL = -0.006 + 0.03 \, PAR \ (75°N-80°N) \qquad (5.3)$$

The goodness of fit is $R^2 = 0.43$. Under given significant level $\alpha = 0.05$, the smaller P value and F values are all convinced and the regressions are significant.

$$CHL = -0.172 + 0.034 PAR \ (70°N-80°N, R^2 = 0.55) \qquad (5.4)$$

$$CHL = -0.208 + 0.034 PAR \ (70°N-75°N, R^2 = 0.44) \qquad (5.5)$$

Next, we do the regression and residual test to see whether they were stationary sequences.

Table 5.7 shows under 5 % significant level, the t-value was −5.0907, which was smaller than the three critical values under 1, 5, and 10 % level. That shows the residual sequence had no unit root, and it was stationary sequence. Hence, CHL and PAR had co-integration relationship, and they also have long-term equilibrium relationship.

5.2.4 The Correlation Analysis Between AOD and PAR

AOD and PAR 10 years' time series is shown in Fig. 5.4 for southern region 70°N–75°N and northern region 75°N–80°N. Generally, AOD was 2 months ahead of PAR. This was confirmed by EViews regression analysis (Pang 2007; Table 5.8).

Table 5.7 shows PAR has good correlation with AOD(−2) in 70°N–75°N. The t test shows the P value is 0.0003, which is less than 0.05. Hence, the correlation is significant. The goodness of fit $R^2 = 0.63$.

After shifting AOD 2 months behind (Table 5.9), the correlation equation is:

$$PAR = -40.0945 + 339.832 AOD(-2)(70°N-75°N) \qquad (5.6)$$

PAR still had better correlation with AOD(−2) in northern region: 75°N–80°N (Table 5.10). The correlation equation is:

$$PAR = 3.558 + 288.7436 AOD(-2)(75°N-80°N) \qquad (5.7)$$

The t test shows the P value is less than 0.001 (Table 5.11). Hence, the correlation was significant. The goodness of fit was $R^2 = 0.37$.

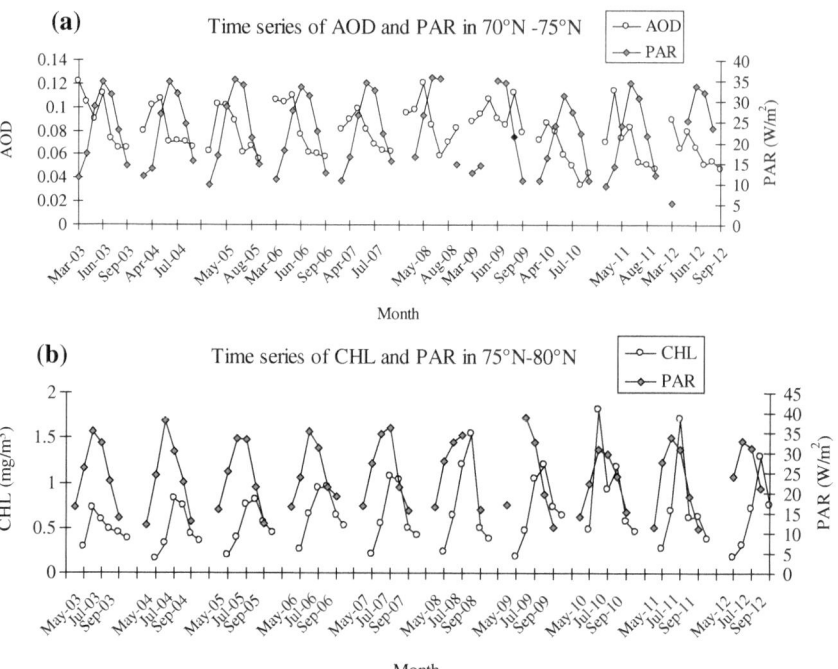

Fig. 5.4 Time series of AOD and PAR for the 10 years 2003–2012 in the subregion **a** 70°N–75°N and **b** 75°N–80°N

Table 5.8 Regression analysis for AOD and PAR in 70°N–75°N

Variable	Coefficient	Std. error	t-Statistic	P value
C	9.8104	9.1925	1.0672	0.2969
AOD(−1)	32.9600	80.7277	0.4083	0.6868
AOD(−2)	288.2451	67.4153	4.2757	0.0003
AOD(−3)	21.5468	84.8235	0.2540	0.8017
AOD(−4)	−155.7919	66.6541	−2.3373	0.0285

Dependent variable: PAR, Method least square
$R^2 = 0.6305$
F-statistic: 9.8099
Prob (F-statistic): 0.00009

Table 5.9 Regression analysis for PAR and AOD(−2) in 70°N–75°N

Variable	Coefficient	Std. error	t-Statistic	P value
C	−4.0945	3.6715	−1.1152	0.2698
AOD(−2)	339.8320	43.5367	7.8057	0.0000

Dependent variable: PAR, Method: least square
$R^2 = 0.5348$
F-statistic: 60.9283
Prob (F-statistic): 0.0000

Table 5.10 Regression analysis for AOD and PAR in 75°N–80°N

Variable	Coefficient	Std. error	t-Statistic	P value
C	11.2945	7.3032	1.5465	0.1351
AOD(−1)	73.5449	107.5236	0.6840	0.5005
AOD(−2)	318.3700	105.9898	3.0038	0.0062
AOD(−3)	−205.5939	126.8522	−1.6207	0.1181

Dependent variable: PAR, Method: least square
$R^2 = 0.4260$
F-statistic: 5.9384
Prob (F-statistic): 0.0035

Table 5.11 Regression analysis for PAR and AOD (−2) in 75°N–80°N

Variable	Coefficient	Std. error	t-Statistic	P value
C	3.5582	5.0592	0.7033	0.4864
AOD(−2)	288.7436	63.3660	4.5568	0.0001

Dependent variable: PAR, Method: least square
$R^2 = 0.3658$
F-statistic: 20.7641
Prob (F-statistic): 0.00006

5.2.5 The Correlation Analysis Among CHL, PAR, and AOD

We have found there were positive correlations between CHL and AOD, CHL and PAR, and PAR and AOD. Now, we still focus on the region 75°N–80°N and see the relationships among CHL, PAR, and AOD. As CHL lagged AOD 2 months behind, CHL and PAR had the same pace. There was no time lag between them. We do the regression analysis for the three (Table 5.12).

The regression equation for CHL, PAR, and AOD is:

$$CHL = -0.163 + 0.03PAR + 20.07AOD(-2)(75°N - 80°N) \qquad (5.8)$$

The goodness fitting $R^2 = 0.58$. Under given level $\alpha = 0.05$, the P value of PAR is smaller than 0.05, but the P value of AOD(−2) is 0.4170, which is greater than 0.05. That means the Eq. (24) was not significant for AOD(−2). This explained that there were significant correlations between any two parameters among CHL, AOD, and PAR. But among the three parameters, the correlation was not significant.

Table 5.12 Regression analysis for CHL, AOD, and PAR (75°N–80°N)

Variable	Coefficient	Std. error	t-Statistic	P value
C	−0.1634	0.1479	−1.1042	0.2757
PAR	0.0282	0.0053	5.2783	0.0000
AOD(−2)	2.0692	2.5248	0.8195	0.4170

Dependent variable: CHL, Method: least square
$R^2 = 0.5785$
F-statistic: 29.5119
Prob (F-statistic): 0.00000

In 70°N–75°N, the regression equation is:

$$CHL = -0.151 + 0.035PAR - 1.02AOD(-2)(75°N-80°N) \tag{5.9}$$

Here, $R^2 = 0.34$, F value is 22.4. The correlation among three was also not significant.

5.3 The Predictions

5.3.1 The Prediction of CHL

EView software is used to do the predictions for CHL for the future several years. We start from the region 75°N–80°N. As CHL is integrated of order one, so we do a first-order differential for CHL (denote it as Y). Table 5.13 shows the unit root-test for Y.

Table 5.13 tells us that Y is the stationary sequence. Next, we draw the Y sequence self-correlation figure to see whether Y is non-white noise sequence. If yes, we set up ARMA model.

Figure 5.5 has two parts: Left column is the self-correlation and second column is the partial correlation. The vertical dotted lines indicate double of the standard deviation. Right-hand side includes 4 columns: The first column is the natural number (indicating the lag order). AC and PAC represent self-correlation and partial correlation coefficient. The last two columns are the Q-stat and correspondent probability.

If a time sequence is white noise sequence, then there is no correlation between the items. Q-stat mainly is for testing whether the sequence is white noise process. It is calculated from the residual self-correlations. If the Q-stat is less than the critical values, then accept the hypothesis that the sequence does not have self-correlation. The Q-stat and correspondent probability (Prob. in Fig. 5.5) show the sequence has correlations. Hence, the sequence is stationary non-white noise sequence.

We try to find the optimal model instead of precise model. The partial self-correlation coefficient approaching 0 after $k = 5$. That means we should choose non-constant item AR(5)

$AR(p)$ model is followed by the following formulae:

$$\begin{cases} x_t = \varphi_0 + \varphi_1 x_{t-1} + \varphi_2 x_{t-2} + \cdots + \varphi_p x_{t-p} + \varepsilon_t \\ \varphi_p \neq 0 \\ E(\varepsilon_t) = 0. \, Var(\varepsilon_t) = \sigma_\varepsilon^2, E(\varepsilon_t \varepsilon_s) = 0, \quad s \neq t \\ Ex_s \varepsilon_t = 0, \quad \forall s < t \end{cases} \tag{5.10}$$

Here, ε is random error term, σ^2 is the variance, and φ is coefficient for each items.

Table 5.13 Unit root-test for Y (first-order differential for CHL). Null hypothesis: Y has unit root

		t-Statistic	Prob.
Augmented dicky-full test statistic		−8.8268	0.0000
Test critical values	1 % level	−3.5402	
	5 % level	−2.9092	
	10 % level	−2.5922	

Date: 04/19/13 Time: 16:59
Sample: 1 70
Included observations: 69

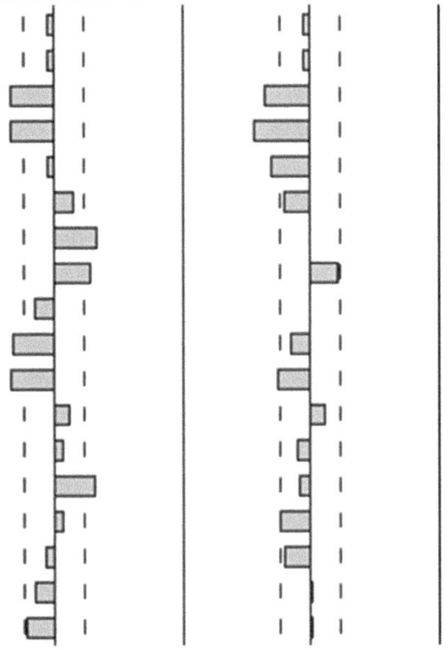

Autocorrelation	Partial Correlation		AC	PAC	Q-Stat	Prob
		1	-0.043	-0.043	0.1348	0.713
		2	-0.054	-0.056	0.3460	0.841
		3	-0.360	-0.366	9.9405	0.019
		4	-0.339	-0.444	18.599	0.001
		5	-0.056	-0.306	18.839	0.002
		6	0.163	-0.205	20.894	0.002
		7	0.354	-0.001	30.790	0.000
		8	0.298	0.226	37.913	0.000
		9	-0.148	0.006	39.709	0.000
		10	-0.323	-0.157	48.349	0.000
		11	-0.358	-0.261	59.153	0.000
		12	0.121	0.127	60.418	0.000
		13	0.072	-0.104	60.873	0.000
		14	0.320	-0.098	70.017	0.000
		15	0.071	-0.242	70.469	0.000
		16	-0.059	-0.199	70.793	0.000
		17	-0.150	0.024	72.905	0.000
		18	-0.235	0.013	78.189	0.000

Fig. 5.5 Self-correlation figure for Y (first-order differential for CHL)

Table 5.14 shows the model fitting is quite good ($R^2 = 0.495$). The equation of the model is:

$$Y_t = -0.491Y_{t-1} - 0.307Y_{t-2} - 0.486Y_{t-3} - 0.645Y_{t-4} - 0.434Y_{t-5}$$
(5.11)
(Where $\varepsilon_t \sim WN(0, \delta^2)$)

Table 5.14 First-order difference sequence for AR(5) model

Variable	Coefficient	Std. error	t-Statistic	P value
AR(1)	-0.4907	0.1171	-4.1891	0.0001
AR(2)	-0.3066	0.1061	-2.8886	0.0054
AR(3)	-0.4859	0.0965	-5.0351	0.0000
AR(4)	-0.6448	0.1074	-6.0014	0.0000
AR(5)	-0.4350	0.1193	-3.6454	0.0006

Dependent variable: Y, Method: least square
$R^2 = 0.4948$
S.E. of regression: 0.33
S.D. dependent variable: 0.5

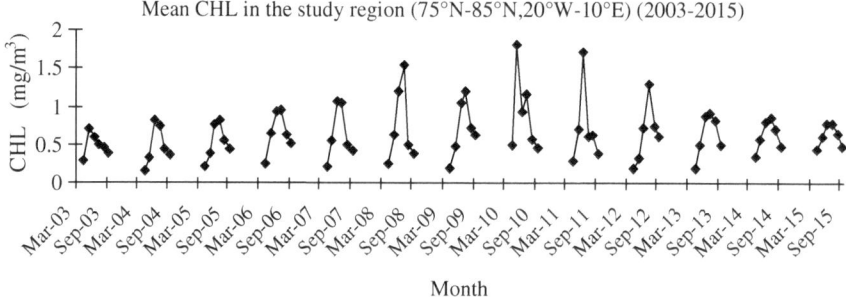

Mean CHL in the study region (75°N-85°N,20°W-10°E) (2003-2015)

Fig. 5.6 CHL predictions for the next 3 years based on the data in 2003–2012 for the region 75°N–80°N (20°W–10°E)

According to this model, we have 10 years data, and future 3 years data could be predicted using ARMA model. According to $Y_t = CHL_t - CHL_{t-1}$, CHL_t is obtained based on previous CHL value at $t-1$. Figure 5.6 is the prediction results for the 2013–2015 for the 3 subregions based on the previous 10 years data. Due to the restriction of the history data, the prediction after 3 years would be out of shape.

5.3.2 The Prediction of AOD

For the prediction of AOD, we still start focus on the region 75°N–80°N for the year 2003–2012 (Chap. 3).

We have obtained the AOD first-order difference sequence Y is a stationary sequence. Now, we need to check whether the first order of difference sequence of AOD is a white noise. Figure 5.7 is the AOD first difference self-correlation and partial correlation figure in subregion 75°N–80°N. The parameter definition is the same as Fig. 5.5.

If the self-correlation of a time series has 0 correlation coefficient, then it has white noise. Q-stat is for checking the white noise. In Fig. 5.7, due to the probability values ("Prob" showing in table) are all significantly smaller than 0.05, Q-stat then all accepts the original hypothesis. On the other hand, the self-correlation was approaching 0; hence, the sequence is stationary non-white noise sequence.

After several fitting processes, AR(7) model is adopted from the Table 5.15. AR(7) has no constant item.

The $R^2 = 0.8626$, and P values are all less than 0.05, and that means the AR(7) model has very good fitting. The equation of the model is as follow:

$$Y_t = -0.534Y_{t-1} - 0.514Y_{t-2} - 0.476Y_{t-3} - 0.444Y_{t-4} - 0.443Y_{t-5}$$
$$- 0.471Y_{t-6} + 0.416Y_{t-7} \quad \left(\text{where } \varepsilon_t \sim \text{WN}(0, \delta^2), R^2 = 0.863 \right) \quad (5.12)$$

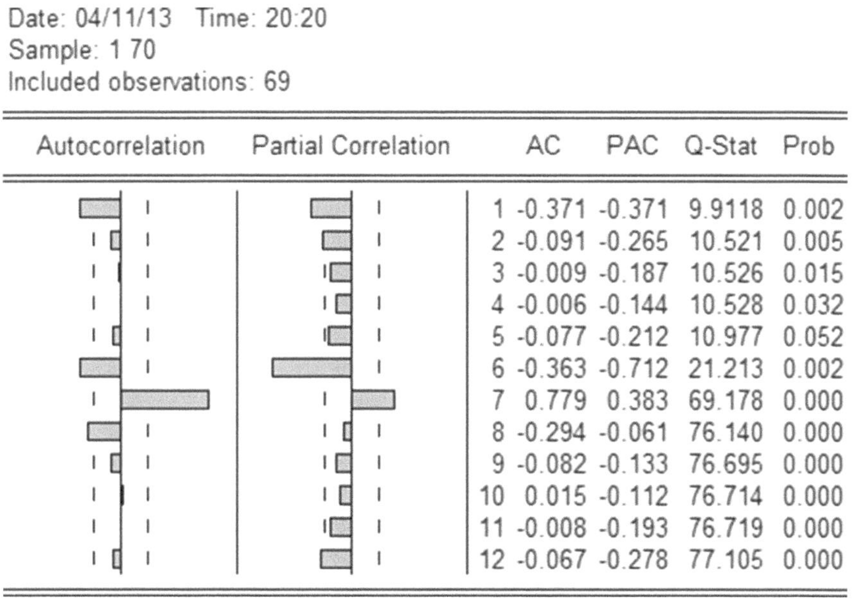

Date: 04/11/13   Time: 20:20
Sample: 1 70
Included observations: 69

Autocorrelation	Partial Correlation		AC	PAC	Q-Stat	Prob
		1	-0.371	-0.371	9.9118	0.002
		2	-0.091	-0.265	10.521	0.005
		3	-0.009	-0.187	10.526	0.015
		4	-0.006	-0.144	10.528	0.032
		5	-0.077	-0.212	10.977	0.052
		6	-0.363	-0.712	21.213	0.002
		7	0.779	0.383	69.178	0.000
		8	-0.294	-0.061	76.140	0.000
		9	-0.082	-0.133	76.695	0.000
		10	0.015	-0.112	76.714	0.000
		11	-0.008	-0.193	76.719	0.000
		12	-0.067	-0.278	77.105	0.000

Fig. 5.7   Self-correlation figure for Y (first-order differential for AOD)

Table 5.15   AR(7) model fitting for Y (first-order differential for AOD)

Variable	Coefficient	Std. error	t-Statistic	P value
AR(1)	−0.5339	0.1221	−4.3740	0.0001
AR(2)	−0.5143	0.1258	−4.0883	0.0001
AR(3)	−0.4760	0.1286	−3.7014	0.0005
AR(4)	−0.4443	0.1278	−3.4769	0.0010
AR(5)	−0.4431	0.1250	−3.5430	0.0008
AR(6)	−0.4710	0.1227	−3.8377	0.0003
AR(7)	0.4164	0.1195	3.4834	0.0010

Dependent variable: Y, Method: least square
$R^2 = 0.8626$
S.E. of regression: 0.0164
S.D. dependent variable: 0.042

Using the AR(7) model, we predict the future 3 years using the previous 10 years data (Fig. 5.8).

5.3.3   The Prediction of PAR

For prediction of PAR in the whole study region, based on the time series of PAR, we obtained the PAR first-order difference sequence Y is a stationary sequence. Next, we need to check whether Y is a white noise. We focus on the study region

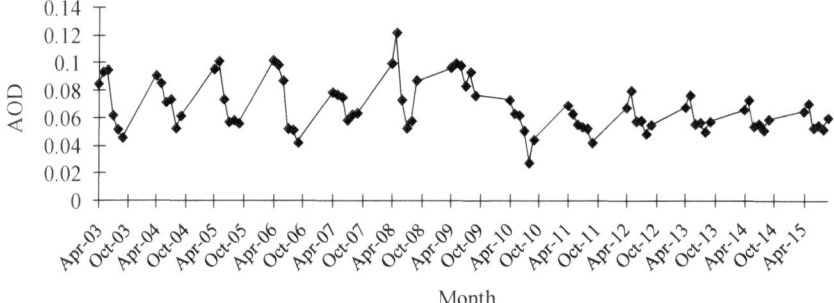

Fig. 5.8 AOD predictions for the next 3 years based on the data in 2003–2012 for the region 75°N–80°N

(20°W–10°E, 65°N–85°N). Next, we draw the Y sequence self-correlation figure to see whether Y is non-white noise sequence? If yes, we set up ARMA model.

Figure 5.9 shows the autocorrelation of PAR first difference sequence has no sign of approaching 0 even after 12 orders. Partial correlation approaching to 0 after $k = 4$. That means we should choose non-constant item AR(4). The PAR first-order difference sequence Y is used to fit the ARMA model.

According to Table 5.16, the regression equation is as follow:

$$Y_t = -0.037124Y_{t-1} - 0.081826Y_{t-2} + -0.416560Y_{t-3} - 0.662566Y_{t-4} \quad (5.13)$$

Date: 03/29/13 Time: 22:52
Sample: 1 70
Included observations: 60

Autocorrelation	Partial Correlation		AC	PAC	Q-Stat	Prob
		1	0.535	0.535	18.072	0.000
		2	-0.140	-0.599	19.338	0.000
		3	-0.687	-0.537	50.104	0.000
		4	-0.737	-0.314	86.211	0.000
		5	-0.222	0.132	89.537	0.000
		6	0.425	0.202	102.00	0.000
		7	0.671	-0.139	133.62	0.000
		8	0.434	-0.095	147.09	0.000
		9	-0.116	-0.027	148.07	0.000
		10	-0.551	0.010	170.65	0.000
		11	-0.500	0.142	189.65	0.000
		12	-0.108	-0.035	190.56	0.000

Fig. 5.9 Self-correlation and partial correlation figure for first-order difference sequence of PAR

Table 5.16 AR(4) model fitting for the first-order sequence of PAR: Y

Variable	Coefficient	Std. error	t-Statistic	P value
AR(1)	−0.0371	0.1254	−0.2960	0.7688
AR(2)	−0.0818	0.1068	−0.7658	0.4483
AR(3)	−0.4166	0.1068	−3.8993	0.0004
AR(4)	−0.6626	0.1259	−5.2625	0.0000

Dependent variable: Y, Method: least square
$R^2 = 0.9453$
S.E. of regression: 1.842
S.D. dependent variable: 7.6

Mean PAR in the study region (65°N-85°N, 20°W-10°E) (2003-2015)

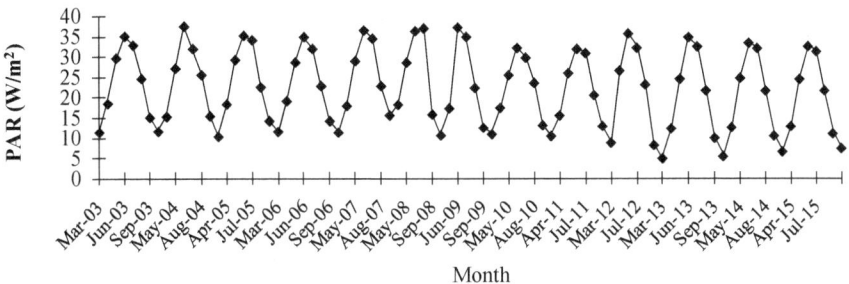

Fig. 5.10 PAR predictions for the next 3 years based on the data in 2003–2012 for the region 75°N–80°N

and here, Y_t is the first difference of PAR sequence, $R^2 = 0.9453$, and it has very good fitting.

Next, according to

$$Y_t = PAR_t - PAR_{t-1} \qquad (5.14)$$

PAR_t values are obtained. The future 3 years prediction is shown in Fig. 5.10.

5.4 Conclusions

The correlation analysis between CHL and AOD, CHL and CLD, AOD and PAR, and PAR and CHL is all studied based on the time series from previous chapters (Chaps. 3, 4 and 5). We focus on the region in 75°N–80°N for correlation and regression analysis. CHL and AOD had positive correlations (0.4–0.5) with AOD 2 months ahead of CHL. CHL and CLD had some correlations, different from year to year. CHL and PAR reached to the similar peak time in June, and they were positively correlated, and correlation coefficient was as high as 0.744 in 70°N–80°N. The goodness of fit was 0.43 in northern region, and the regressions were significant.

AOD was 2 months ahead of PAR, and the goodness of fit were 0.63 and 0.42 for 70°N–75°N and 75°N–80°N, respectively.

The correlation among CHL, AOD, and PAR has studied. There was a correlation among CHL, AOD(-2), and PAR, but the correlation was not significant in the study region.

Prediction for CHL, AOD, and PAR is done finally. CHL, AOD, and PAR all had non-smooth sequences, with their first-order differential sequences were smooth non-white noise time series. Therefore, the future 3 years predictions have been done using ARMA model (with AR(5) for CHL, AR(7) for AOD, and AR(4) for PAR).

References

Carmack, E. C., & Macdonald, R. W. (2002). Oceanography of the canadian shelf of the beaufort sea: a setting for marine life. *Arctic, 55*(Suppl. 1), 29–45.

Gabric, A. J., Cropp, R., Ayers, G. P., McTaubsh, G., & Braddock, R. (2002). Coupling between cycles of phytoplankton biomass and aerosol optical depth as derived from SeaWiFS time series in the Subantarctic Southern Ocean. *Geophysical Research Letters, 29*(7), 1112–1146.

Gabric, A. J., Qu, B., et al. (2005). The simulated response of dimethylsulfide production in the Arctic Ocean to global warming. *Tellus B, 57*(5), 391–403.

Harrison, W. G., & Cota, G. F. (1990). Primary production in polar waters: relation to nutrient availability. *Polar Research, 10*, 87–104.

Horner, R. A. (1985). *Ecology of sea-ice microalgae. Sea-ice biota* (pp. 83–103). Boca Raton: CRC Press.

McCree, K. J. (1981). Photosynthetically active radiation. *Encyclopedia of Plant Physiology, 12A* (pp. 41–55). Berlin: Springer.

Pang, H. (2007). *Econometrics* (pp. 265–284). Beijing: Science Publishing Press.

Qu, B., Gabric, A. J., Lu, H., & Lin, D. (2014). Spike in phytoplankton biomass in Greenland Sea during 2009 and the correlations among chlorophyll-a, aerosol optical depth and ice cover. *Chinese Journal of Oceanology and Limnology, 32*(2), 241–254.

Spies, A., Brockmann, U. H., & Kattner, G. (1988). Nutrient regimes in the marginal ice zone of the Greenland Sea in summer. *Marine Ecology Progress Series, 47*, 195–204.

Chapter 6
Conclusions and Discussions

6.1 Conclusions

Greenland Sea is an important part of Arctic Ocean with its special geophysical locations. The melting of Greenland glaciers and sea ice in Greenland Sea had great impact on the local ecosystems. We investigated the sea-ice cover in Greenland Sea and its relationships of the followings:

(1) Phytoplankton biomass (CHL) and North Atlantic Oscillation (NAO);
(2) Aerosol optical depth (AOD) and cloud cover (CLD);
(3) Photosynthetically active radiation (PAR) and sea surface temperature (SST).

Our research region is in the Greenland Sea in 20°W–10°E, 65–85°N during the recent 10 years period 2003–2012. Remote sense satellite data were used for the research. The distributions of all parameters are listed in the study region. Furthermore, we divided into four different 5° zonal apart subregions. The distributions are compared. The peak times are calculated, and lag regression analysis using both spreadsheet and Eviews software are carried out for two parameters relationship. General agreements are arrived for each scenario. We also did the three parameters regression analysis. The predictions for future 3 years for CHL, AOD, and PAR are done in the end.

Results shows CHL with ICE, AOD, PAR, and NAO all have long-term equilibrium relationship. The melting ice (MI) played a significant role on promoting the growth of CHL, increasing of AOD. ICE had more impact on AOD than CLD to AOD. ICE had less impact on PAR although ICE was 2 month ahead of PAR.

We are more focused on the northern region (75°N–80°N) where ice melted more. The reasons of unusual higher CHL located in the north of the study region (near 79°N–80°N) was studied. The enhanced water-column stability and more MI (more runoff irons), less salinity were all the reasons. The driving forces of unusual earlier and higher peaks of CHL in 2010 were also studied. The higher wind speed in spring, and wind direction changed from southeast to southwest direction, brought more MI in the northern region. Relative mild SST and lower PAR profile in year 2010 plus negative NAO (indicating more ice melting in the year) all favored the growth of phytoplankton biomass.

B. Qu, *The Impact of Melting Ice on the Ecosystems in Greenland Sea*,
SpringerBriefs in Environmental Science, DOI 10.1007/978-3-642-54498-9_6

Finally, we predicted future three years (2013–2015) time series for CHL, AOD, and PAR. ARMA model was used, and finally, three models (AR(5), AR(7), and AR(4)) were used for predicting CHL, AOD, and PAR, respectively.

Our research on sea-ice impact on Arctic ecosystem is only at its primary stage. More researches are expected to fill the gap and correct the errors.

The book gives general outline of ecosystem in Greenland Sea and how the ice impacts the local ecosystems. The book also provides valuable statistical methods on correlations analysis and predicting the future ecosystems.

6.2 Discussions

6.2.1 The Role of Sea Ice

The loosing of Arctic sea ice is causing a great concern worldwide. It would have great impact on the Arctic ecosystem and also would influence the global climate. The sea ice serves essential habitat in Arctic ecosystem, such as the source for prey and sources for life cycles of many organisms. The loss of sea ice would introduce species from lower latitudes and would change the primary productions (Moline et al. 2008). The melting of sea ice would increase the stratification of water columns and influence the ecological processes and interactions between different water bodies. On the other hand, sea ice acts as a physical barrier to prevent the direct exchange between ocean and the atmosphere. It could keep warm ocean water in winter. The loss of ice would lead to the heat loss to atmosphere and increase the coming solar radiation from the air to sea. The loss of ice would also alter the ocean currents. The lower salinity of surface water would favor the primary productions.

Different from Antarctic Ocean which has iron-limited water body, Arctic Ocean is supplied with trace elements from river runoff and from the atmospheric dust deposition. The formation of ice is mixed with sediment particles. The MI would bring source of iron to primary productions. However, different from Antarctic, Arctic sea ice is nitrate limited (Smetacek and Nicol 2005).

6.2.2 More About Arctic Amplification

Arctic amplification has numerous causes as we mentioned early on. The key factor of albedo feedback has direct and indirect effects on the warming. The direct effect on temperature rising during sunlit season (summer) and indirect effects on the seasonally lagged effect are associated with the early summer ice melting and heat gain in the surface mixed layer (Curry et al. 1995). All the processes of causing Arctic amplification work reversely to force the Arctic cooling.

In contrast, the melting of ice cover in southern ocean does not have strong impact on the albedo feedback. On the other hand, the surface heat in the southern ocean is rapidly removed from the surface and does not have much influence on the ice cover (Stroeve et al. 2007). Arctic amplification is expected to be stronger in the next coming decades. It will impact the Arctic and beyond. We face a challenge: How to slow the amplification speed and even turn the part of warming to cooling? This work helped us understand more about the ecosystem in Arctic Ocean, how ice impacts the Arctic Ocean, and how things might change.

6.2.3 Accuracy of the Satellite Data

Due to the remote research region in Arctic Ocean, the only available data are satellite data. How accuracy of the data is? This question could be raised all the time. Here are our explanations.

CHL concentration (represents phytoplankton biomass data) is from satellite MODIS level-3 aqua data. The aqua data we used are more stable than terra data. SeaWiFS was merged into MODIS after 2002. NASA team had carried a polarization correction for MODIS productions. The correction procedures had good agreement with the MODIS aqua water, leaving radiance time series with the data from another. The accuracy is improved. The benefit is especially from Arctic region (for 412 nm band, the 50 % decrease of the ratio during northern hemisphere winter has been removed). MODIS team did calibrations for regional differences between MODIS (terra and aqua) with SeaWiFS (Lwn's) (Bailey and Werdell 2006). They used in situ measurements as ground truth to do calibration. The area covered the most of global area (up to northern most latitude). The validation includes the measurement scale and the time window considerations.

However, there is still an unavoidable error could occur for solar zenith angles more than $70°$ and view angles more than $45°$ (Bailey and Werdell 2006). The satellite data are more accurate with the presence of sunlight and the absence of clouds and sea ice. Hence, the data are more accurate in summer time comparing to the early spring and winter time. It is reported that the satellite CHL underestimated the concentration in the field by a factor of 1.4 when using in suit CHL data averaged over the penetration depth. Hence, the depth-varying CHL data are expected for more accurate valuations rather than only surface CHL data. The adjusted CHL satellite algorithms are expected for polar region to overcome the underestimation situations.

PAR algorithm in SeaWiFS used 412- to 670-nm radiances, and it assumed the effect of cloud could be decoupled from the clear atmosphere. Hence, the error is unavoidable from the satellite PAR data. The relative biases for PAR retrieval are 4.6 % for all sky and 2.9 % for clear sky (Su et al. 2007).

6.2.4 Global Warming or Cooling?

The global warming is caused by changing of human activities. The oceanic CO_2 absorption capacity decreases, while oceanic CO_2 burden increases. Hence, the global warming is unavoided. It is not well known that on the other hand, atmospheric aerosols have a cooling effect on global temperature. In the large ocean such as Arctic, dimethylsulfide (DMS) emission is significant. As DMS is a key compound in the global sulfur cycle and its fluxes to the atmosphere influence atmospheric acidity, at the same time, it changes cloud formation and hence cools the earth temperature (Charles et al. 1987; Malin 1997). Marine phytoplankton biomass is a main source of precursor for DMS in the ocean. DMS sea-to-air flux could contribute to the formation of cloud condensation nuclei (CCN) and affect the radiative budget and rate of warming (Charlson et al. 1987). As DMS is the main sulfur released during the decay of ocean biota, how much does the ocean biota produced DMS offset the greenhouse warming? Gabric et al. (2013) used an atmospheric GCM with incorporated sulfur cycle, coupled to a mixed-layer ("q-flux") ocean, to estimate the climatic response to a prescribed meridionally variable change in zonal DMS flux. They found that the vertically integrated global mean DMS concentration increases by about 41 % after perturbation. Global mean surface temperature would decrease by 0.6 °K after perturbation. This perturbation on DMS flux leads to a mean surface temperature decrease in the southern hemisphere of around 0.8 °K, comparing with a decrease of 0.4 °K in the northern hemisphere. The results are not significant; however, it cannot be ignored. With the decreasing of ice cover, there will be an increase on phytoplankton biomass, which would lead to more DMS flux in the Arctic Ocean. The cooling offset in Arctic could be amplified in the future.

6.2.5 Further Research

The changes of sea ice would impact ice algal communities. Climate change has also resulted in a great loss of sea ice in Arctic Ocean. Researchers suggested that mild climate change might increase ice algal production, while increased stratification would reduce phytoplankton productions (Tedesco et al. 2012). Higher temperature could increase the grazing rate and nutrients regeneration (Melnikov et al. 2009). Reducing of sea-ice area eventually would decrease the annual phytoplankton production contributed by sea ice. The ecosystem would be changed by decreasing of ice, while animals lost winter food sources and hard to survive. The reasons causing blooms in spring and early summer are complicated. Apart from ice melting, stable water stratification and the ocean current-related forcing cannot be ignored. The melting of ice could also have negative impact on the climate by increasing the surface temperature. The timing of melting is important to the ice algae blooms. The earlier melting trends lead earlier blooms. They will lead to increasing rate of grazing and nutrients regenerations. The ecosystem would be changed accordingly. All of those combinations made the project of ice impact to climate more challenging.

It is also a debate for how much the MI has positive impact to the phytoplankton biomass, hence cooling the climate?

There are still much more detailed researches need to put forward. Comparing to other region, the deeper advective mixing occurs in Greenland Sea and the vertical profile of CHL is worth to explore. Cherkasheva et al. (2014) has started this work.

Greenland Sea is in a special geophysical position in Arctic Ocean. Understanding Greenland Sea-related climate variability and changes is a challenge. It is influenced by local physical drivers and also by poleward head transport from North Atlantic Ocean and other oceans. The interaction of the movements is related to the surrounding oceans. Hence, it is a long-term project.

Sea-ice zone could be a source of atmospheric CO_2 in winter and a sink in summer and fall (Miller et al. 2011; Else et al. 2008). Anthropogenic changes in Arctic region are pronounced, and its impact will be the largest. Melting ponds could transmit light to underlying ocean and stimulate the under-ice primary productivity. Under-ice blooms may be significant larger than surface blooms, and they may contribute significantly to primary production in the region. Measurements and modeling of primary production under ice requires immediate attention. A sound understanding of the sea-ice melting and its effect on the earth system in the large scale is highly necessary.

References

Bailey, S. W., & Werdell, P. J. (2006). A multi-sensor approach for the on-orbit validation of ocean color satellite data products. *Remote Sensing of Environment, 102*, 12–23.

Charlson, R. J., Lovelock, J. E., Andreae, M. O., & Warren, S. G. (1987). Oceanic phytoplankton, atmospheric sulphur, cloud albedo and climate. *Nature, 326*, 655–661.

Cherkasheva, A., Nöthig, E. M., Bauerfeind, E., Melsheimer, C., & Bracher, A. (2014). From the chlorophyll-a in the surface layer to its vertical profile: a Greenland Sea relationship for satellite applications. *Ocean Science, 9*, 431–445.

Curry, J. A., Schramm, J., & Ebert, E. E. (1995). On the sea-ice albedo climate feedback mechanism. *Journal of Climate, 8*, 240–247.

Else, B. G. T., Papakyriakou, T. N., Yackel, J. J., & Granskog, M. A. (2008). Observations of sea surface fCO2 and estimated air-sea CO2 fluxes in the Hudson Bay region (Canada) during the open-water season. *Journal of Geophysical Research (Oceans), 113*(8), C08026. doi:10.1 029/2007JC004389.

Gabric, A. J., Qu, B., Rotstayn, L., & Shephard, J. M. (2013). Global simulations of the impact on contemporary climate of a perturbation to the sea-to-air flux of dimethylsulphide. *Australian Meteorology and Oceanographic Journal, 63*, 365–376.

Malin, G. (1997). Sulphur, climate and the microbial maze. *Nature, 387*, 857–859.

Melnikov, I. A., Nihoul, J. C. J., Kostianoy, A. G. (2009). Recent sea-ice ecosystem in the Arctic Ocean: a review. In J. C. J. Nihoul & A. G. Kostianoy (Eds.), *Influence of climate change on the changing Arctic and sub-Arctic conditions* (pp. 57–71). NATO Sci. Peace Secur. Ser. C. Dordrecht: Springer.

Miller, L. A., Papakyriakou, T. N., Collins, R. E., Deming, J. W., Ehn, J. K., Macdonald, R. W., et al. (2011). Carbon dynamics in sea-ice: A winter flux time series. *Journal of Geophysical Research: Oceans, 116*, C02028.

Moline, M. A., Karnovsky, N. J., Brown, Z., Divoky, G. J., Frazer, T. K., Jacoby, C. A., et al. (2008). High latitude changes in ice dynamics and their impact on Polar Marine ecosystems. *Annals of the New York Academy of Sciences, 1134*, 267–319.

Smetacek, V., & Nicol, S. (2005). Polar ocean ecosystems in a changing world. *Nature, 437*, 362–368.

Stroeve, J., Holland, M. M., Meier, W., Scambos, T., & Serreze, M. (2007). Arctic sea-ice decline: Faster than forecast. *Geophysical Research Letters, 34*, L09501.

Su, W., Charlock, T. P., Rose, F. G., & Rutan, D. (2007). Photosynthetically active radiation from Clouds and the Earth's Radiant Energy System (CERES) products. *Journal of geophysical research, 112*(G02022), 1–11. doi:10.1029/2006JG000290.

Tedesco, L., Vichi, M., & Thomas, D. N. (2012). Process studies on the ecological coupling between sea-ice algae and phytoplankton. *Ecological Modelling, 226*, 120–138.